*Body Leaping Backward*

# BODY

# LEAPING

# BACKWARD

*Memoir of a Delinquent Girlhood*

## MAUREEN STANTON

HOUGHTON MIFFLIN HARCOURT
BOSTON • NEW YORK • 2019

For information about permission to reproduce selections from this book,
write to trade.permissions@hmhco.com or to Permissions, Houghton Mifflin Harcourt
Publishing Company, 3 Park Avenue, 19th Floor, New York, NY 10016.

hmhco.com

*Library of Congress Cataloging-in-Publication Data*
Names: Stanton, Maureen, author.
Title: Body leaping backward : memoir of a delinquent girlhood / Maureen Stanton.
Description: Boston : Houghton Mifflin Harcourt, 2019. | Includes
bibliographical references.
Identifiers: LCCN 2018033154 (print) | LCCN 2018052035 (ebook) | ISBN
9781328900364 (ebook) | ISBN 9781328900234 (hardback)
Subjects: LCSH: Stanton, Maureen, author. | Female juvenile
delinquents—Massachusetts—Walpole—Biography. | Drug abuse and
crime—Massachusetts—Walpole. | Walpole (Mass.)—Social conditions. |
BISAC: BIOGRAPHY & AUTOBIOGRAPHY / Women. | BIOGRAPHY &
AUTOBIOGRAPHY / Cultural Heritage. | SOCIAL SCIENCE / Criminology.
Classification: LCC HV6046 (ebook) | LCC HV6046 .S73 2019 (print) | DDC
364.36092 [B]—dc23
LC record available at https://lccn.loc.gov/2018033154

Book design by Michaela Sullivan

Printed in the United States of America
DOC 10 9 8 7 6 5 4 3 2 1

Lines from "Sodomy" (from *Hair*): Lyrics by James Rado and Gerome Ragni. Music by Galt
MacDermot. Copyright © 1966, 1967, 1968, 1970 (copyrights renewed) by James Rado, Gerome
Ragni, Galt MacDermot, Nat Shapiro and EMI U Catalog, Inc. All rights administered by EMI
U Catalog, Inc. (publishing) and Alfred Music Publishing (print). All rights reserved. Used by
permission of Alfred Publishing, LLC. Lines from "Don't Let It Bring You Down" and "Only
Love Can Break Your Heart": Words and music by Neil Young. Copyright © 1970 by Broken
Arrow Music Corporation. Copyright renewed. All rights reserved. Used by permission.
Reprinted by permission of Hal Leonard LLC. Lines from "It's All Behind You": Words and
music by Andy Pratt. Copyright © 1973 by EMI April Music Inc. Copyright renewed. All
rights administered by Sony/ATV Music Publishing LLC, 424 Church Street, Suite 1200,
Nashville, TN 37219. International copyright secured. All rights reserved. Reprinted by
permission of Hal Leonard LLC.

Photographs used by permission of the author.

FOR MY MOTHER

*The first eighteen years really shape you forever. It's like a glass of water filled with mud. You can pour clear water in until it appears clear, but there's still mud there.*

— Bruce Springsteen

# CONTENTS

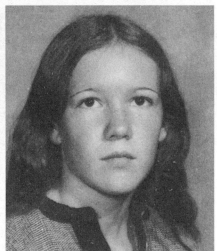

*Top, from left:* Patrick, Susan, my mother holding Michael, Joanne, my father holding Barbie, Sally, and me, 1970. *Bottom left:* My mother graduating from nursing school, 1975. *Bottom right:* Me at fifteen years old.

*Top left:* Me, Barbie, Sally, and Joanne in Stanton, California, 1976. *Top right:* Michael, Sally, my father, Barbie, Joanne, Patrick, and Sue, with me patched at right. *Bottom left:* Me at my work-study job, 1977. *Bottom right:* Me at Muir Woods, California, 1978.

*You Can't Even Get Out*

IF YOU GROW UP ON THE SEACOAST, YOU LEARN TO SWIM, TO navigate choppy water. The flatlands of the Midwest teach you about spaciousness and its possibilities, the safety of sameness but also tedium. In a factory town you learn about labor and time clocks. Growing up in Walpole, Massachusetts, home to the state's maximum-security prison, I learned about good and bad, about being inside or outside, about escape.

In the mid-1960s, when my siblings and I were little (six of us at the time), if my mother was driving past Walpole State Prison, she would slow the station wagon to a crawl along the shoulder of the two-lane road. "See that place?" she'd say, her head lowered to peer out the window, our faces pressed to the glass. "If you misbehave, you'll end up in there." My mother couldn't put too fine a point on her lesson. "See the fence? It goes all the way around."

It *was* strange — that huge building with massive white walls surrounded by dense cedar forest, like something out of a fairy tale. I thought the walls looked like the papier-mâché we made in first-grade art class — that same eggshell color. Around the perimeter was a chainlink fence topped with a curl of razor wire, like our Slinky toy stretched on its side. "If you're not good, that's where you'll end up," my mother would say. "You can't even get out. Take a good look. Imagine spending your whole life in there."

Who knows where we were going on those drives — maybe to the discount clothing store in nearby Plainville. Whatever our destination, it wasn't urgent enough to prevent my mother from taking advantage of the prison as a behavior-modification tool, a gigantic real-life object lesson. For my mother, the prison was a boon to parenting, an inescapable specter of destiny writ large in black and white, like the stripes of the jailbird in Monopoly. Once we saw the prison, once it lived in our imaginations, my mother could conjure its symbolism to discipline us. If my sisters and brother and I bickered, if we kicked and punched each other or aggravated each other by mere proximity, crammed in the backseat of the car ("Mom, Sally's *breathing* on me," or "Mom, Joanne won't stop *staring* at me"), my mother yanked the car to the side of the road or craned her neck toward the backseat. "If you don't behave, I'll put you in Walpole Prison!"

A decade later, my mother stands in our kitchen about to drive to the Registry of Motor Vehicles. She is dressed in nice pants and a blouse, her dark brown hair pinned in a twist, mascara highlighting her nearly black eyes, lipstick outlining her movie-star smile. In her late thirties, she is a little thick in the waist after her seventh child, but still pretty and petite — high heels raise her to five feet tall. She looks like who she is — a thirtysomething suburban housewife, not a person about to commit a felony.

I'm fifteen and at least I look like who I've become — a druggie, a delinquent. The hems of my ratty jeans are frayed from dragging on the ground, my faded dungaree jacket is too big, my hair is pulled straight and parted in the middle; the start of a vertical frown line divides my brow, mark of an angry young woman. So much has changed in a decade in my family, in the country. Categories have shifted, boundaries blurred. Who are the good guys? The bad guys? What's right and wrong anymore? Nothing is as clear and defined as it was in those hopeful early days of the 1960s when my mother drilled into her children a strict moral code, simple lessons made concrete by the concrete walls of the prison. Good and bad, inside and outside, the walls a solid, reliable boundary between the town and the prison that shared a name: Walpole.

My mother slides into the driver's seat of her car, onto the pillow that allows her to see over the steering wheel. Her pocketbook on the passenger seat holds forged papers to transfer ownership of a stolen camper. "Keep your fingers crossed," she says as she puts the car in reverse. "I could wind up in jail." Her words have a similar cautionary tone as when we drove by the prison years before, but the message is the opposite, and not abstract. She's not warning against bad behavior but against getting caught.

*Body Leaping Backward*

# 1

## *Here We Are Living*

I HAVE A MEMORY THAT, DECADES OLD, STILL MAKES MY HEART ACHE, a filmstrip that ticks through my mind's eye like this: Late spring 1965, the morning sunny and warm as my family visits our house-to-be in Walpole, a small town twenty miles south of Boston. A sign in a vacant lot reads PINE TREE ESTATES, with a faded map of plots. *Estate* is an aspiration, an exaggeration of the modest framed-in houses. We are among the first families here, and we feel like pilgrims, settlers. The builder offers a couple of models — split-level ranches, gambrels — so there is the appearance of diversity, but the pattern repeats unimaginatively until the pavement abruptly ends at pine forest. A sign at the top of the street reads, inauspiciously, DEAD END.

My mother, Clarissa, and father, Patrick, my four sisters — Sue and Sally, who are older than me, Joanne and Barbie, who are younger — and my brother, Patrick, also younger, are here. My mother is twenty-six, my father thirty, and the six of us are seven, six, five, four, two, and one — Barbie, swaddled in my mother's arms, is why we need a bigger house. (Mikey, the seventh, will be born in five years.) My mother wears a sheath skirt and waist jacket, pumps, her dark hair pulled back, with spit curls like earrings. We've gone to church and now have driven over to our house being built.

My father is smiling; I can see the diastema in his front teeth, a gap I'll inherit and close with braces in another seven years. He wears a suit and tie, like when he leaves for work each morning. I don't know what my father's job is, and the next year in first grade when Miss Hanson asks us to draw a picture of our father at work, I grow anxious. In the mid-1960s the nuclear family is largely intact, so it's safe for Miss Hanson to assume that all kids have fathers at home, that those fathers go to work each day. *Nuclear family* is a curious term, coined in 1946, the year after our country dropped two nuclear bombs on Japan. *Nuclear* connotes cohesion — protons and electrons revolving around the nucleus — but also an explosion: something bound together, something torn apart.

In class that day, my page blank, I begin to cry. Miss Hanson kneels by my desk. "Does he work in an office?" I recall a visit to my father's office, many desks in a large room, swivel chairs, typewriters. "I think he is a secretary," I tell Miss Hanson, who looks doubtful. "Are you sure?" But I'm certain and I get busy drawing a typewriter, happy that I've realized my father is that long, important word. When I bring my picture home, my mother says, "Your father is not a *secretary*." The way she says *secretary*, I know it's lesser, but I don't know why. My father brings home stacks of three-by-eight-inch rectangles in heavy manila stock and gives them to us for coloring. These papers have tiny square holes punched through them in no particular order, like windows in an office building. I will be in my twenties before I vaguely understand my father's work as a systems analyst, when computers were the size of rooms and programmed with punch cards.

In the shell of our house we wander through rooms framed with two-by-fours, passing through walls like spirits. I breathe in a sharp pleasant smell, sawdust and wood, but I won't know its name until shop class in ninth grade, the astringent scent of pine lumber. After we've seen everything, we linger in this outline of a house, trying it on like new winter coats, admiring ourselves, the whole of us, measuring the hope we feel in the air, taking breaths of it. We are expanding to occupy this space. This is the age of expansion: the population, the economy, geopolitics, space. To Americans, everything is big and growing bigger.

Our family, too: big and growing bigger. The backyard is covered with tons of rocks — pea gravel for a leach field. That night in my dreams the rocks become M&M candies, an alchemy of the imagination, a dream that signifies sweetness, abundance. Beyond the gravel is a steep hill that in four years Nancy Morris will sled down, and break her collarbone, though for now Nancy Morris is a girl I don't know living in Cedar Rapids, Iowa, and nothing is yet broken.

Twenty or so families lived on our dead-end street, as if on a peninsula; there was only one way in, one way out. The families were young, all but one white, with a baby or two born every year, families at the beginning of their promising lives. The houses were painted in primary colors — red, yellow, blue, green, or sometimes chocolate-brown or white. The driveways were paved. The lawns were trim, yards landscaped with shrubbery. I was glad we had shrubs, which thwarted the kidnappers that my Aunt Barbara promised would carry me away in a huge sack if I didn't behave, shrubs literally a hedge against intrusion.

We were told to watch out for kidnappers, a word that confused me — someone who *nabs* kids when they are napping. There was a kidnapper in *Chitty Chitty Bang Bang*, the Child Catcher, who rode around in a truck that hid a jail cell. It was unclear what the kidnapper did with the kids once he nabbed them. In bed at night when I couldn't sleep, I'd watch the window for a silhouette of the kidnapper, who I imagined looked like Andy Capp, that comic-strip character from the Sunday paper, a snub of cigarette glued to his lip and no eyes, just a nose poking out from his cap. When I studied the shadows through translucent curtains, it was Andy Capp I waited for in fear.

In spring and summer parents chatted in the street at dusk while kids played kickball or dodgeball. At twilight mothers stuck their heads out their front doors and screamed the names of their children, calling them home. A few mothers used whistles to summon their kids: two shrill blows for the Stewarts, three short sharp tones for the Murphys. I wished my mother had a signal for us, even though the whistles seemed horrible, like whistling for a dog. That seemed to happen often when I was a child — I envied or desired something that also horrified me.

After the streetlights sputtered on, the older kids played Flashlight Tag, hide-and-seek in the dark. Home base was called "gools." I don't know the origin of the word, but my father used it in his games. *Gaol* is "jail" in Gaelic, so perhaps gools was something my father heard from his immigrant parents or the many Irish in Dorchester, his Boston neighborhood. To us, gools was home, and was always our front steps. When I was nine, ten, eleven, it seemed I'd never tire of flashlight tag on summer nights, of venturing farther from gools, skulking from tree to tree to stay hidden, and then at a carefully timed moment casting myself into the night and racing to gools, to safety. The chase left me breathless, as if I were running for my life.

At the end of our street the pavement stopped abruptly at a bluff, and from its edge we'd leap ten feet below into the forgiving sand. Beyond the sandpit, in the soft duff beneath towering white pines, my sister Sally and Sherry Stewart and I built forts, though we didn't build them as my brother did, nailing lumber to trees. Rather we outlined rooms on the pine-needle carpet with sticks and rocks: kitchen, living room, bedrooms, sometimes napping on the cushiony moss beneath the pines and balsam firs, like nymphs, listening to the creaks and groans of branches rubbing in the wind. The woods felt like home to me.

One day I picked a beautiful lady's slipper that grew in the understory, its blossom like a pink silk purse. I presented this exquisite gift to my mother, but she said never *never* pick them because they were rare and it was against the law. We would have to pay a $50 fine. Sherry Stewart picked the lady's slippers anyway; she didn't care about the law. That was the first time I knew of anyone intentionally breaking a law, and that person was a child.

On one side of us lived the Petersons, who'd moved there after their oldest child, Mark, was killed by a drunk driver. The accident left another son, Brad, with a jagged scar above one eye and a plate in his head, my mother told me. I spent an inordinate amount of time try-ing to imagine the plate in Brad's head, envisioning a tiny flowered tea plate like my mother kept in a hutch. Later I heard it was a *metal* plate and I wondered if it felt cold, like the headache you get from eating ice cream too fast.

Across the street were the Gibsons, Connie and Arthur and their daughter, Peggy, her long uncombed hair in knots, "like rat's nests," which her mother, unlike mine, did not painfully yank out, which in my mind was clearly a failing. Arthur Gibson wore droopy green Dick-ies on weekends as he tended plots of vegetables and flower beds, his glasses slipping off his pointy nose as if gravity were getting the better of him. Mrs. Gibson always wore snap-down housedresses and open-toed mule slippers with white ankle socks. The Gibsons' house was al-ways astonishingly and, it seemed to me, unapologetically messy, Mrs. Gibson in the den every afternoon, the blue flicker of television tuned to soap operas.

The Wagners lived next door, Eugene and Joan, and their children, Judy and Billy. The Wagners' yard was perfect: no toys lying around like in our yard, no wrappers or deflated balls or balloon scraps, or pieces of wood with protruding nails, my brother Patrick's construc-tion projects. Only forty feet separated our house from the Wagners', but their split-level was situated downslope from our house, which en-abled us to look down upon their lives, to see everything.

Billy Wagner was four when his family moved next door, but already he was bad. When Billy was punished, he was confined to his house, sometimes with yard privileges. Mostly Billy sat at the boundary be-tween our properties, clearly visible by the Wagners' manicured lawn and our shaggy dandelion-infested grass, which my father avoided cut-ting in favor of playing tennis on Saturday mornings.

*Property* was a word I learned early, particularly in its relationship to trespass. In the woods behind our house, signs read NO TRESPASS-ING — like in church, those who trespass against us. Our properties were like tiny kingdoms over which we reigned. When neighbor kids squabbled, we yelled, "Get off my property!" One day Doreen Randall —the girl whose father was the drunk in front of Tee-T's downtown, the girl who lived in the house with asbestos siding that I thought of as a "tarpaper shack" like I'd read about in some book—wandered into the Gibsons' yard where we were playing on the swing set, her smile exposing rotted teeth. Peggy Gibson said, "Doreen. You are not *allowed* on our property." I watched Doreen's face crumple as she walked away: those skinny, dirty legs and scuffed patent leather shoes that we wore only on Sundays. I felt the heat of shame for remaining silent as I pumped my legs higher. But what could I do? It wasn't my property.

Billy Wagner sat at the edge of his property as if it were the edge of flat earth and watched the neighbor kids playing. When we came near him, he'd say, "Will lou play with me?" (He couldn't pronounce the letter *y*.) Billy looked like a child-sized white Sidney Poitier, with a round face and full mouth, bright brown eyes. To play with Billy we'd have to enter the Wagners' yard, but Mrs. Wagner was always watching. We glimpsed her figure sieved through the screened porch, or her face in the kitchen window. We knew she was watching; she knew we'd transgress. Transgression was the mission of childhood: to push against boundaries. "Let's go exploring," we'd say, which meant turning over rocks and poking into vernal pools, breaching new neighborhoods like Oak Street Extension and Ridley Avenue, riding our bikes alongside the train tracks farther and farther.

Nobody played with Billy, so he punished us. "Caca-head," he'd yell. "Sticks and stones will break your bones, but names will never hurt you," my mother said, but that was a lie; the names people called you, the labels, broke through your skin, got under it. Sometimes Billy made a mad dash a few feet over the invisible line and attempted to poke one of us with a stick. Our defense was simple: *I'll tell your mother.* Billy would whimper, beg us not to tell.

One day I sat inside my mother's old navy-blue Ford, which we called the Beetle. I loved to play inside the car, because it was toasty warm and smelled like heated plastic, but mostly because it was quiet and private. Nobody knew where I was, and I took pleasure hiding in plain sight if my mother or one of my sisters called my name, searching for me. That day I saw Mrs. Wagner yell out the front door, "William!" *Billy* was the sleeping dog of trouble; *William* meant trouble as fait accompli. Billy's face got that horrible, wide-eyed, close-to-tears look, as if he were breaking. I saw Billy cowering behind the rhododendron next to the chimney. I saw Billy hiding and I saw Mrs. Wagner on the front steps, the whole inevitable circumstance unfolding before me.

I watched from the car like I was watching a drive-in movie, spellbound, with a sick, anxious feeling in my stomach. When Mrs. Wagner found Billy, she hauled him along with a firm grip on his skinny upper arm. I could almost feel the clench of her fingers. She pushed him with her hand on the back of his neck, Billy half running to her long strides. Just before they reached the garage door, she fully extended her arm and struck him, as if she could not restrain herself five seconds longer until they were inside, where I couldn't see, where nobody could see. I heard the *thwack* of her hitting him again, his high-pitched crying, the door slamming.

Billy was always bad, and Judy, his older sister, was always good. They were the real-life versions of Goofus and Gallant, my favorite section of *Highlights* magazine, parallel drawings of the good boy and bad boy elucidating right from wrong behavior. Judy had silky white-blond hair, which she usually wore in a stretchy headband, sapphire-blue eyes, and a longish face. In my memory, Eugene and Joan and Judy Wagner all had long faces, chins slightly stretched as if made of Silly Putty, but now I think their faces seemed long because they never smiled.

Perhaps I was obsessed with Billy Wagner because I recognized in him a shadow of myself. I, too, was bad, it seemed. At the dinner table I sat on my father's left, and somehow on many nights I did something that irritated him. Barbie, the youngest for six years, until Mikey was born, remembers watching the sequence unfold — my speaking out, being "fresh," reaching for something with my "long hands." Barbie knew I was going to be slapped and it made her anxious. She wondered why I wouldn't just be quiet. *Shut up.* This dinner-table sequence happened so often that my mother confessed to me years later that she told my father, "You've *got* to stop hitting Maureen."

I can't remember specific incidents of freshness at the dinner table, only that I was compelled to speak my mind and apparently it could not be slapped out of me. *Loudmouth. Long hands. Dunderhead. Fresh.* I was "more outspoken" than the other kids, my mother told me. "Like what did I say?" I asked her. "There's too many to remember," she said. "You're not supposed to sass your mother. You don't sass your father." What lesson was my father trying to teach me, slapping sense into me? Do not take, grab, reach. Do not defend yourself. Do not "talk back" or be a "smart-mouth." It was my nature — critical, inquisitive, outspoken — that he was trying to squelch, the slaps a slow steady pressure like water over stone. I was a girl who acted as girls should not; there was a need to smooth the rough edges.

Is it possible I was born with a tendency to freshness? One of our annual family portraits, taken when I was three, is revealing. My mother and father sit side by side, five children arranged around them. I'm standing on the bench next to my father, my pudgy right arm behind his head, my fingers in rabbit ears, a prank. Was I headstrong? Have poor impulse control? Or did I possess ordinary desire that had to be tamped down, the expectation of girls to be polite, to acquiesce, to be good? I was "nosy." When my mother was angry, she said, "Jesus H. Christ," but she wouldn't tell me what the *H* stood for. It was just Jesus's middle name. *Why can't I know that?* I sensed taboo like a fragrance on the wind, like faint smoke from the leaf piles neighborhood fathers burned in the fall.

One day in the bathroom wastebasket I spied an interesting waxy blue bag. When I unwrapped it, to my horror I found a huge white bandage soaked with blood. I ran to my mother. "Mom, who was hurt?" She looked puzzled. "Nobody." I showed her the evidence. "Mind your own business," she said, "and stay out of the trash." I have a history of seeing for myself, taking that step over the line. When I was five, I stood behind Doug Frick as he aimed his metal horseshoe at the post thirty feet away. He told me to back up, but when he turned around, I crept closer for a better view. He swung the horseshoe and whacked me in the forehead. Blood dripped down my face as my mother drove me to Dr. DeRoma, whose crude stitch left a small scar over my left eyebrow.

Here was a glimpse of the girl I was, my desire too much, a lesson I'd have to learn again and again. One Thanksgiving I asked my mother to put extra barley in the turkey soup because I loved barley. She assured me that she'd added plenty, but I didn't trust her. When nobody was around, I climbed onto the kitchen counter, my head grazing the plaster ceiling. I reached into the cabinet for the barley and dumped some into the pot, a huge warped aluminum cauldron filled with bubbling broth. Somehow my foot bumped the pot, which sat off-kilter on the electric burner, and it crashed to the floor, steaming broth spreading across the linoleum, the turkey carcass surfing to the middle.

I was embarrassed about ruining the soup, but more so for being caught wanting.

Sometimes my freshness was for a good cause. One afternoon Patrick threw a rock at Billy Wagner, who'd been pelting my brother with pebbles in the hope that Patrick would play with him. Billy ran crying into his house, and soon Mrs. Wagner walked onto their screened-in porch. "Patrick," she said. "Were you throwing rocks at Billy?" Billy smirked as he wound himself around his mother's legs; he knew that he'd re-channeled her enormous wrath onto someone else. If Patrick told the truth, Billy's mother would beat him. If Patrick protected Billy, then Patrick would be in trouble. Patrick said nothing. He was probably six, so I was nine.

My mother stepped onto our porch as Mrs. Wagner said, "Answer me, young man!" The mounting tension of the confrontation, the hypocrisy of Mrs. Wagner scolding Patrick, boiled up in me, boiled over. "Shut up!" I yelled at Mrs. Wagner. I believed that I was rescuing my brother, defending his honor. I thought my mother would be *proud* of me. I sensed that she didn't like Mrs. Wagner, but I didn't yet comprehend that you were not supposed to be honest, that you should suppress your real thoughts and feelings and be polite. *Shut up,* I'd said to an adult (the admonishment I so often received from adults). My mother was furious. She dragged me into the bathroom and shoved a bar of soap in my mouth, a too-literal punishment.

I was only trying to heed my mother's advice after Frank Richards pushed me down one day when I was seven and I ran sniffling to her as she folded a mound of laundry in the cellar. "You've got to learn to fight your own battles," she said. I remember the shift in my psyche that day — that I could no longer count on my mother to extricate me from trouble or come to my defense. I had to fight my own battles, but I was not supposed to be fresh. No wonder I was confused.

In fourth grade I ordered a paperback from the Scholastic Book Club, *Sacco and Vanzetti, Murderers or Murdered?* The two men on the cover — Italian immigrants, one with a sad scraggly mustache draping his mouth — were found guilty of murder, which troubled me. *They had alibis!* Sacco and Vanzetti were executed in 1927 in spite of protests worldwide, especially in Boston. That book haunted me — innocent men put to death, a wrong that could never be righted.

Sacco and Vanzetti had been held in Charlestown State Prison in Boston, built in 1805, one of the oldest prisons in the country. By the 1950s the antiquated Charlestown prison needed to be replaced, so the Department of Corrections built a new state-of-the-art prison in Walpole, an institution meant to be progressive, to rehabilitate inmates, 97 percent of whom would be released eventually. "Most men and women in the state's prisons want to return to society as better citizens," said Russell G. Oswald, Massachusetts's corrections commissioner then, who became corrections commissioner in New York in 1971, just months before the Attica Prison riot that left twenty-nine prisoners and nine hostages dead.

Before the new Walpole State Prison opened in 1956, some of the walls had already cracked because of "insufficient thickness," and locks were found to be easily popped open with a penknife. Once the problems had been fixed, five hundred or so prisoners were relocated to Walpole and the decrepit Charlestown State Prison was razed. For forty years Walpole was the state's only maximum-security prison —

security level six — where the most notorious, most dangerous criminals were sent.

To get to Walpole Prison from my house, you drove down a steep hill
past a cemetery, where on summer days I rode my bike between the
rows of headstones. At the bottom of the hill was the blacksmith's shop,
a dark sooty garage where a stooped white-haired man fixed our bikes
when our mother ran over them with her station wagon. In spite of her
constant reminders, we dropped our bicycles in the driveway behind
her car, and so, because we'd been warned, "it serves you right" that
your bike was crumpled. My mother didn't say this maliciously but as
if it were a law of nature; we got what we deserved. My father took our
bikes to the blacksmith, who for $5, under the blue flame of his blow-
torch, hammered the rims straight.

At the blacksmith's shop, you turned right on Main Street to downtown, past Dixon's Five and Dime, Friendly's restaurant, the A&P, past
Joe's Produce, which always smelled of overripe tomatoes, with fruit
flies swarming the wooden bins, past Watson's candy store, where
hairnetted ladies in white uniforms hand-dipped chocolates, past the
skinny street, which we begged my mother to take for the novelty —
only one car could pass at a time and the steps to the buildings emptied
right onto the asphalt. The skinny street was where the Wests lived, six
children with wide-set eyes and limp black hair. Whenever you passed
through downtown, you almost always saw a Westie. Main Street became Route 1A as buildings thinned and you entered a stretch of forest.
Emerging from the greenery like Oz was the prison on the right, its
white walls a blank canvas.

The prison reminded me of the castle we visited in France, in the
Loire Valley, where we lived for two years when my father worked on
computer systems for the military. Originating in the ninth century,
the Château de Blois was expanded over the centuries to 564 rooms.
During the tour, Joanne, who was two, began to fuss, so my parents
held us back, my mother jostling Joanne in her arms. When we tried
to catch up with the group, we found ourselves alone, the tour leader

having forgotten about the American family with four little girls in matching home-sewn dresses. We walked down a winding staircase to a vaulted antechamber with stone walls and two enormous wooden doors, which were locked. My father knocked, but nobody answered. He knocked and knocked, then pounded his fists against the door, yelling, kicking the door, all four girls crying now. It was as if we were in some horrible fairy tale, our lonely shouts reverberating in the gloomy castle, darkness falling; trapped. A half-hour later, a guard heard my father and freed us.

For years after I had dreams of wandering in the castle, the damp stone walls, those tall wooden doors. When my mother said, "If you're not good, I'll put you in Walpole Prison," I envisioned the castle in France.

Walpole Prison was not the only, the first, or the largest prison in the area. A mile away and closer to my house was Norfolk Prison, a medium-security prison for 1,500 or so men. Combined, Walpole and Norfolk Prisons housed more than half of the male convicts in the state then, an epicenter of incarceration and punishment. Originally called Norfolk Prison Colony in 1927, Norfolk was world-renowned as a "prison of the future," with dormitory-style housing, not cells with bars. In the early years, inmates tended cows, pigs, and chickens and grew vegetables to feed their brethren. They learned carpentry, plumbing, and welding or took college classes. Norfolk Prison offered music and theater, and there was an extraordinary debate team that regularly defeated teams from Harvard, Princeton, MIT.

One member of Norfolk Prison's debate team was Malcolm Little, later known as Malcolm X. At twenty-one and addicted to drugs (he smoked pot and snorted cocaine daily, he wrote in his memoir), Little was caught with loot from armed robberies around Boston. He was convicted and sentenced to ten years, even though two years was the average term for first-time offenders of those crimes. At Norfolk, Little joined the debate team, a "baptism in public speaking," he wrote, where he found his voice.

Walpole Prison, opened thirty years after Norfolk, was also a bea-
con of hope at first. "All the cells are spacious and airy, with good light-
ing and plumbing," said Oswald, the corrections commissioner. "There
are three chapels, a gym, a three-chair barber shop, and the 'better-
behaved' inmates may be allowed to use the guards' bowling alleys."
Even the inmates were optimistic. "There's a new spirit here," said
John Flaherty, serving thirty-one to forty years. "Here we are living. All
we could do in Charlestown was die." Articles about the prison were
mostly positive, like one about prisoners building wooden models of
the SS *Hope,* a floating medical center treating poor people around the
world, and folding five thousand milk-carton collection boxes to raise
money for the ship, like the boxes we assembled in school to collect
donations for UNICEF. Inmates rebuilt used toys and bikes donated by
Walpole's citizens and gave the toys to orphans at Christmas. "Behind
those cold cement walls there are some warm hearts," editorialized the
*Walpole Times,* the town newspaper.

After the prison opened, the majority of Walpole citizens surveyed
felt "a responsibility to help make the prison a part of the community,"
and they lived up to that belief, the town showing "a fine spirit of interest in its new neighbor," the newspaper reported. The United Church
Bible Group and the Toastmasters "lent a hand" to demonstrate the
"good will of our community." Walpole residents donated books to set
up a prison library, and the Footlighters, an amateur theater group,
staged productions with the inmate troupe, the Masquers. A Walpole
mother with the Footlighters said the inmates "treated us in a most
gentlemanly way, and then some." Methodist Church members toured
the new prison (separate tours for men and women), and Jim Regan, a
former Walpole High School football star and the prison's new athletic
director, appealed to Walpole residents for donations — "a catcher's
mask, mitt, bat, or a football."

My music teacher, Mr. Willey, helped inmates stage a variety show
in the prison for five hundred guests, including Walpole's prominent
citizens. "There were dancing teams, acrobats, crooners, comedy acts,
a saxophonist. A man who used to play organ in some of the most mag-
nificent churches in the country, now a Walpole inmate, was called

back for several encores." I wonder if Mr. Willey liked working with prisoners, grown men, more than with schoolchildren, who didn't appreciate his efforts. Mr. Willey seemed ancient when I had him in elementary school, silver hair rippling across his head. He'd stand at the front of the room in a suit and tie, conducting us through patriotic songs — *Over hill, over dale, / as we hit the dusty trail / and those caissons go rolling along* — What was a caisson? Nobody explained — *then it's hi! hi! hee! / in the field artillery / shout out your numbers loud and strong.* We especially liked shouting *HI! HI! HEE!,* which pleased Mr. Willey, who mistook our volume for civic pride. We always sang Mr. Willey's favorite, "My Country, 'Tis of Thee," the mimeographed lyrics held in our hands, his eyeglasses steaming up from emotion as we shout-sang the rousing finale, *Le-et free-dom ring.*

When I was growing up, nobody called the prison by its long awkward title — Massachusetts Correctional Institution at Walpole. They'd say "Walpole Prison" or just "Walpole," the prison synonymous with the town, as if we all lived in one huge barbed-wire pen. I felt a perverse pride that Walpole Prison was in our town, named after our town, a pride of ownership; it was ours, for better or for worse. The prison made us famous, notorious. Don't mess with Walpole. We had a state champion football team, a winning field hockey team, and the maximum-security prison.

The prison was like a watermark on the town, seeping into our psyches in subtle ways. One Halloween my mother sewed prison attire for herself and my father for a costume party — not guards' uniforms, but inmates'. She cut long-sleeved shirts and pajama-like pants from black-and-white-striped fabric, using as her reference the cartoon jailbird from Monopoly. She stenciled numbers on the backs of the shirts and spray-painted black two Styrofoam balls, larger than softballs, which she and my father chained to their ankles with a construction paper chain. With their matching striped pillbox hats, they were the hit of the party, my mother told me the next day.

The next year, Sally, Joanne, and I devised our own crime-themed costumes for the Girls Missionary Guild Halloween party. We'd been recruited into the GMG by our neighbor Sherry Stewart, whose mother held weeklong Bible camp in summer, which we attended mainly for the candy. Sally, Joanne, and I were the Odd Squad, after our favorite TV show, *The Mod Squad,* about three teenagers who avoided jail by working undercover for the police. Pete had been arrested for stealing a car, Linc for rioting in a fictionalized version of the Watts race riots, Julie for running away. The show debuted in 1968 after the end of the Hays Code, which decreed that criminal acts onscreen must be punished: "The sympathy of the audience should never be thrown to the side of crime, wrongdoing, evil or sin." In the first season of *The Mod Squad,* Julie, Linc, and Pete busted a gang of thieves working in cahoots with crooked cops, a plot which presaged the 1970s around the corner, when it became increasingly difficult to discern the good guys from the bad guys.

In our Odd Squad, Sally, the tallest, was Linc. Joanne was Pete, decked out in striped bell-bottom pants and a fake mustache. I was Julie, in a minidress and love beads. Our costumes won first prize, a King James Bible, which I read one afternoon like an engrossing novel filled with stories of thievery, murder, sin.


On Sundays we attended Mass at Blessed Sacrament, a beautiful, imposing neo-Gothic brick church with wide granite steps leading to two tall wooden doors, the apostles carved into their panels. Inside, past the narthex, where you dipped your finger in holy water, stained glass windows along the side walls depicted the stations of the cross. Sometimes I took the framed print of Jesus off our wall at home and stared at him —his brow dripping blood, hollows beneath his eyes—trying to feel something more than pity, more than curiosity. My memory of church is bathed in blood, the carpet on the chancel and altar, the stained glass, the wine, the wounds. Beneath the vaulted ceilings of Blessed Sacrament, religion was infused with a somber sense of shame and guilt: the stain of sins, the murder of Jesus.

My father always led us to a middle pew, never the front, which seemed overly pious, nor the rear, which seemed disrespectful. The middle pew perhaps symbolized my father's growing disillusion with Catholicism. I'd catch his slight eye roll as the Richards family, gloved and hatted, strolled to the front pew every Sunday. Except for singing alleluia, Mass was intolerable, broken up by the ritual of placing coins in a straw basket attached to a long pole, shoved into each row by an impatient old man. I fantasized grabbing the money from the basket as it passed under my nose, all those dollars people foolishly gave away.

When we'd been slothful and had missed the last service at Blessed Sacrament, we raced to St. Jude's, a Catholic church just past Norfolk Prison, for the last Mass of the day, at 11:15 a.m., Saint Jude the patron saint of lost causes and desperate cases. St. Jude's was to Norfolk Prison as Blessed Sacrament was to Walpole Prison: less strict. St. Jude's didn't even look like a Catholic church, with its sand-colored facade, tiny windows, and plain decor, no bloody Jesus anywhere.

In catechism we learned the Ten Commandments. Thou shall not kill, thou shall not steal—these I understood. But thou shalt not covet thy neighbor's wife? I thought of the neighbor ladies, Mrs. Peterson with her raspy cigarette voice, or Mrs. Gibson, with her soap-opera-watching ways, or mean Mrs. Wagner. Why would I covet those wives? There was one sin I committed, though it wasn't a commandment: pride. One day in first grade, Joanne and I knocked on the Gibsons' door, calling for Peggy to play. When Mrs. Gibson opened the door, I told her that I'd earned straight A's on my report card. At home, Joanne told my mother and my mother scolded me for bragging.

On Saturday mornings my father took us to confession in Blessed Sacrament's annex, a miniature version of the church with a dozen pews, a small altar, and confessional booths with velvet curtains to draw for privacy. At your turn you entered the booth, kneeled, and spoke into the screen, behind which you could see the shadow of Father Cummings, could hear his wheezy breath. You'd say, "Forgive me, Father, for I have sinned. It has been ___ days since my last confession." Ideally it should be a number of days, not decades. Tracking a week of sins was difficult. I couldn't remember all that had happened, so mostly I confessed generic sins, acts I was likely to commit, a kind of preemptive confession: stealing a quarter off my father's dresser, hitting my sister, lying to my mother.

I was honest once, telling the priest I'd done nothing wrong. I don't remember what the priest said, but I sensed that I was not going to get out of that uncomfortable, stuffy booth and off my suffering knees until I offered something. From then on, instead of the ridiculous task of tracking actual sins, I began to confess made-up sins. I sometimes wished I had bigger and better sins to confess, but I was not yet an accomplished liar; I was an apprentice liar. I tried to be honest, to say what I thought, but I was learning, albeit slowly, that honesty resulted in a slap, a scolding, soap on my tongue, or pepper. Penance.

After the first couple of times saying penance — five Hail Marys and two Our Fathers, for example — I realized that no one checked whether you said the prayers, or said them right. The church used the honor system. Why did the priest trust someone he knew to be a sinner, someone who regularly — every Saturday! — arrived with a fresh litany of sins? Did he really think this person, this incorrigible recidivist person, was to be trusted carrying out her own penance? I must not have believed in God, or else I'd have worried that he was watching. Could I have been agnostic at the age of seven, eight? Instead of repenting, I kneeled at the altar for a certain amount of time, studying the intricate frescoes, Jesus on that cross, nails pounded into his living flesh, the ignominy of hanging alongside two thieves, his slow, agonizing, cotton-mouthed, miserable death.

My mother was deeply religious, beyond being a practicing Catholic — she'd perfected it. Lent, Ash Wednesday, Palm Sunday — I observed the holy days and rituals without understanding them. Why did the priest press his ashy thumb on my forehead and leave his mark? I didn't know and didn't care, but my mother lived her faith. She taught my third-grade catechism class and cooked and delivered meals for the elderly and shut-ins. One year for the church bazaar she constructed a Barbie dollhouse from an appliance box, laboring after dinner each night for weeks. The house had a bedroom, living room, bathroom, kitchen, the rooms wallpapered with decorative contact paper, with tiny ruffled curtains she sewed for the windows. She fashioned a couch and an easy chair, cutting forms from cardboard and upholstering each piece, the seats lifting to reveal a storage place for doll clothes. The house was like the dioramas we made in school, though in comparison to our shoeboxes, my mother's house was a mansion.

Night after night as I watched my mother work on the house, I begged her not to donate it to Blessed Sacrament church's annual bazaar and white elephant sale.

"But Mom," I'd say, "why are you going to just give it away?"

"It's for the poor," she mumbled through straight pins clamped in her perfect white teeth. I imagined "the poor" as Doreen Randall, the girl Peggy Gibson had banished from her property.

I thought that through sheer persistence I would convince my mother not to donate the house and instead give it to someone who'd truly appreciate it: me. I believed my relentless pleading would change her mind, even as we transported the dollhouse to the church sale and set it on a table next to a raffle jar. When the raffle was drawn, I watched the parents of some little girl cart the house away. *She doesn't look poor,* I thought. Nor did she look thrilled, I observed. She didn't see the beauty of the house, only the old appliance box it had been.

My mother had her own set of cardinal rules, like a tailored version of the seven sins. *Never throw stones. Don't lie. Don't steal. Never ever say the* n *word. Don't cheat. Don't talk back. Don't hit.* Only adults could hit. At first my parents' system of discipline was orderly. When we were still spanking age — up until five or six, after which there was the occasional empty threat: *You're not too old for a spanking* — during the day my mother calculated our misdeeds on an abacus in her mind, and when my father came home from work she presented him with the sum. My father sat on the piano bench and we lined up to receive a spanking, like the orphans receiving gruel in *Oliver Twist*. One night I stuffed a book down the back of my pants, something I probably saw Alfalfa do on *The Little Rascals*. Did it hurt my father's hand? Is that why he didn't think me clever and instead doled out an extra ration of punishment?

My parents abandoned the system by the time I was seven, perhaps because our misdeeds were too numerous to track. Probably the system simply broke down, entropy being the inevitable end of all systems. Then punishment took the opposite form, not regimented but random. When we fought, my mother punished everyone involved. She wasn't interested in our pleas for justice, that the punishment was *not fair* because *she started it*. On the two- or three-hour drives to our annual weeklong vacation in some rented cottage in New England, if we fought in the backseat, my father threatened to pull over and give us a spanking. He never did this, though, not once. Instead, while driving sixty miles an hour, steering with his left hand, he'd flail wildly with his right hand into the backseat, striking innocent and guilty alike. From this random, hit-or-miss punishment, I learned that it didn't pay to be good.

From early on the world seemed violent; my eye was drawn to violence, my psyche disturbed by it. There was Billy Wagner next door, and then on my first day of first grade, after my mother dropped me off (Joanne, Patrick, and Barbie still at home), I stood in line outside Fisher Elementary waiting to enter. When the bell rang, Jill Fletcher wailed and clung to her mother's leg, snot bubbling from her nose. Her mother spanked her, *whack whack whack,* her arm swinging at her daughter's behind. The more Jill cried, the more her mother scolded and hit. Jill's mother seemed embarrassed to have this messy daughter, who was finally pried off her leg and who sat sniffling and hiccupping the entire morning, the only girl besides me who cried on the first day of school, we two crybabies. My mother told me later that I was her only child who cried on the first day of school for many years.

Neil Kelleher lived near Barnes's corner store, situated between our neighborhood and the projects, a collection of twenty or so small ranch houses. Neil had black hair and close-set brown eyes and overlapping front teeth. One day I stood in Barnes's parking lot unwrapping a

Bonomo Turkish Taffy. I wasn't supposed to be eating candy, especially since I'd bought it with quarters I'd stolen off my father's dresser, so I had to finish the taffy there in the parking lot. I saw Mrs. Kelleher step outside. "NEIL!" she called, then waited. "NEEEEE-IL!"

The Turkish taffy pulled at my teeth as I watched Mrs. Kelleher in her driveway, curlers in her hair, hard spiky hot electric curlers like my mother used, the suburban housewife's crown of thorns. She cupped her hands to her mouth, her voice shrill, and finally Neil careened into his driveway on his banana bike. When he skidded to a stop his mother pulled him off the bike, which clattered to the pavement. As she swatted him he fell to the ground, crouching, covering his head with his hands as she beat him with her shoe, thwacking him across his back.

I stood riveted, thinking of how she'd removed her shoe, how odd that was, turning an ordinary object into a tool to beat someone. Did she fear that if her own hand touched the flesh of her son's back in rage she might realize what she was doing? That she might connect the back she was pounding with the back she'd patted when Neil was a baby? I felt sick from the gooey taffy I shouldn't have been eating. I couldn't seem to swallow or get the taffy out of my mouth, the awful repetitive thump of a shoe on a boy's back like a stick against a rug, the sickening sweet taffy gluing my mouth shut.

Neil Kelleher was Patrick's friend through high school, until Patrick took off for California and then Hawaii, as far as he could get from Walpole without leaving the country, and Neil became trapped in Walpole, in prison on a drug conviction.


Growing up, I was almost never alone. The house was filled with those others, my mother and father, my sisters and brothers; between them I filled a space, a shape of an identity extruded through the crowd. "There is no such thing as a single human being, pure and simple, unmixed with other human beings," writes sociologist Nancy Chodorow. Each of our personalities, she says, is "a company of many," a composite formed of "never-ending influences and exchanges between ourselves and others." In that house I was hardened into being. We were close in age, like a litter, with slight differences like markings. In every direction I pushed, someone was there. The two sisters above me, Susan and Sally, "Irish twins" born ten months apart, had straight black hair, plump lips. The two sisters below me, Joanne and Barbie, were taller and thinner, with straight black hair and small heart-shaped mouths. I was the middle of five girls, but different, with lighter curly hair, a mouth somewhere between the plump lips of Sue and Sally and the cupid's-bow lips of Joanne and Barbie. My two brothers, Patrick and Michael, both younger than me, had my mother's big dark-brown eyes and long eyelashes.

Sue was extroverted, a people person, a leader in school. Sally, the sister I was closest to in childhood — she's a year minus five days older — was reticent and moody, artistic and bookish. I was like Sue and Sally both, a hybrid. Joanne was quick-witted, always smiling and happy, but hesitant, not a risk-taker, and stubborn. Barbie was detail-oriented and intuitive, with a touch of the sixth sense — nobody could beat her at Concentration, that card game where you find matching pairs from memory. Patrick was bright and handsome and athletic, like my father, but he was at times overwhelmed by five sisters (four older), especially if we clustered, sprawled on my mother's bed, poring through boxes of hand-me-downs from our cousins, one of us modeling the outlandishly unfashionable clothes, all of us laughing hysterically. Patrick would see us gathered and flee. He longed for a brother, and so my mother had her only planned pregnancy, delivering on her promise to Patrick. Mikey was six years younger than the next youngest, Barbie. I was ten when Mikey was born, so he was like a living doll, a chubby happy baby, ten pounds at birth.

My mother strived to treat us equally, dividing M&M's into mathematically even portions. Equality was an abiding principle impossible to follow, but she tried, each of us an enforcer (at least for our own benefit), complaining if we were slighted in the slightest way — an M&M could launch a tiny war. Certain things presented problems: only one shotgun seat in the car, only two legs on a roast chicken. One had to look out for one's interests. When my mother came home from shopping, I helped her put away the groceries with the ulterior motive of squirreling away food for the lean days at the end of the month, the days of peanut butter crackers for lunch. For years I stockpiled cans of Progresso lentil soup in my underwear drawer. (Patrick hid the beef barley.)

When there was nothing else to eat, I'd plunder a loaf of Wonder Bread, which helped build strong bodies in twelve ways — the girl in the TV ad growing before my eyes as if she'd eaten a magic mushroom. I'd tunnel my hand through the loaf like an earth-boring mole, grabbing the soft center, wrist-deep in the yellow, red, and blue polka-dot plastic bag. I'd compress the bread in my palm, making a dense golf ball of dough, grayish from my dirty hands. Then I bit the dough ball, satisfied by the tracks my teeth left. My mother hated this ruination of the loaf, but she could not catch the culprit, the girl who stole her family's bread. Having siblings taught me to steal, to hoard, to share.


On weeknights when we were small, my siblings and I would race out of the house to greet my father when he came home from work, one kid sitting on each of his shoes, clinging to his calves as he walked into the house weighted down by children. After he changed out of his suit, the aroma of roast chicken or pork chops filling the house, he played boogie-woogies on the piano, an upright Wurlitzer, songs like "The Worry Man Blues" or "Bill Bailey Rag," as we danced around the living room. He played songs that were popular during his childhood, like "Barney Google" with his *goo-goo-googa-ly eyes,* which I associated with Joanne, with her big round black eyes and eyelashes so long and thick they clumped with sleep sand.

Some songs were so sad — like "500 Miles Away from Home," about a lost soul *cold and tired and all alone* — that I felt a hollow in my chest, like when a tooth fell out and there was just an empty hole and tenderness. And Tom Dooley, a poor boy who must hang down his head and cry, for tomorrow he'll be hanged from a white oak tree, even though he claimed innocence — Tom Dooley, always forever about to die. My favorite was an Irish ballad, "The Wild Colonial Boy," just sixteen when he left home, robbing the rich to help the poor, shot to death by police.

I had a little crush on the Wild Colonial Boy, who I imagined as the actor Jack Wild, who played the artful dodger in *Oliver!,* the first movie I saw in a cinema, in 1968. I cried at the horrific scene of Billy Sikes clubbing Nancy under the bridge, embarrassed to be crying in public, even though it was a dark theater; my father was surprised when the lights came up, and he put his arm around me. "Faucets," my family teased, because I cried easily. The Wild Colonial Boy and Tom Dooley and that lonesome stranger so far from home — they worried me as I sang along with my father at the piano, both of us off-key, his body swaying, my heart wrenched open.

At night as we watched *Ed Sullivan* or *Family Affair* or *Please Don't Eat the Daisies,* my mother summoned us one by one to lie on the kitchen counter and hang our heads in the sink as she vigorously shampooed our hair, deaf to complaints. On that same counter she slapped slices of bologna onto an assembly line of buttered bread for school lunches, forgetting sometimes that I preferred mustard, and on those days I couldn't eat the slimy bologna made slicker with slabs of butter.

Until we got too heavy, at bedtime my father carried us two at a time from the basement TV room up two sets of stairs. We gathered on someone's bed to hear my father spin stories about Sammy Beetle-bug, Danny Dragonfly, and Irving Spider, who battled the Oak Street gang, my father wearing different hats to act out characters. In bed we shouted to him, "Bring me some cold, cold, cold, cold water," all of us chanting, a cacophony, making sure we wouldn't be forgotten, which was always the danger. After my father made the rounds with the same cup refilled many times, he'd play the piano downstairs to lull us to sleep, someone calling for another song: "I'm not asleep yet!"

Sometimes Joanne, my roommate, and I sang in bed, at least until she drifted off. I'd lie awake wondering when I would fall asleep, knowing that as long as I was wondering this, I wouldn't, not knowing how to stop waiting for sleep to envelop me like a wave at the beach. In the dark I'd trace my finger on the embossed ballerinas on my wallpaper as if I were reading Braille like Helen Keller, one figure in arabesque, another en pointe.

One summer night when I couldn't sleep I heard voices outside. Standing at my window, I watched my parents walk across the street to the McGraths' for a late-night dip in their pool, my father with a towel wrapped around his waist like when he stepped out of the shower. I concluded that they were skinny-dipping. I stood for a long time trying to catch a glimpse of a silvery naked body, eavesdropping in the hope of snatching some exotic adult word that I could take back to bed and mull over, wonder about its portentous meaning, like the word *seduction,* which I'd seen on a gag gift someone gave my parents, a miniature four-way street sign showing the stages of coupledom: seduction, love, marriage, divorce. I didn't understand that *divorce* was the punch line. Or the word *abandon,* which I read in my father's *Time* magazine, a story about a mother who dropped her children at a park and never returned. I asked my mother what *abandon* meant, but I couldn't fathom it. "But *why* did she leave them?" Until then I thought that trouble came from outside, like the kidnappers I imagined lurked outside my window.


We traveled as a pack, moving through time and space together, connected by an invisible tether, bound by rituals of pleasure. In winter, legs wrapped around each other, we tobogganed the giant hill behind the high school. We skated on Memorial Pond downtown, my father kneeling to tighten everyone's laces one at a time. Sometimes on Friday nights we ate pizza and fried clams at Tee-T's, cramming into a booth on the restaurant side, the jukebox twanging through the thin wall that separated us from the garrulous men smoking and drinking on the bar side. On summer weekends we went to the Braintree twin drive-in theater, dressed in pajamas at twilight, playing on the swings with other kids in pajamas, like being awake in a dream. When the sky darkened and a cartoon flickered on the screen, we raced to our car to watch

*True Grit* or *Chitty Chitty Bang Bang* or *Planet of the Apes.* Once in a while, because of someone's poor planning, the second screen of the twin drive-in showed a film meant for adults. I'd crane my neck to see enormous close-ups of lips and faces pressed together, flesh projected large onto the silver screen, an alternative narrative to the one allowed by my parents, a silent sensual story.

On Sundays after church we sprawled belly-down on the living room carpet eating glazed doughnuts and reading the funnies in the *Boston Globe,* my father in his easy chair with the boring sections, my mother in the kitchen cooking a roast. Out of the blue my father would say, "Let's go to the beach!" We'd change into bathing suits and flip-flops, stuff towels into our beach bags. My mother would wrap the whole roast beef, still warm in its Pyrex baking dish, and lug it across the sand as if it were an eighth child. Sitting on blankets at Duxbury Beach, we ate sandy roast beef sandwiches on white bread pinkened by blood. Years later I asked my mother why she insisted on portaging a roast to the beach. "Was it so important to have Sunday dinner?"

"That wasn't my doing," she said. "I'd be cooking a roast like every Sunday, then halfway through the day your father would get an idea in his head to go to the beach. So I packed up the roast and took it with us." I glimpsed the woman my mother was, a woman happy to put her husband and children first, to accommodate and please no matter how impractical, a woman I could never be.

My father tried to instill some culture in us, taking us to museums and shows, like a Clancy Brothers concert: four Irishmen in off-white cable-knit wool sweaters, looking exactly as they did on the album covers at home. One year he took the four older girls to a play, *You're a Good Man, Charlie Brown,* at the Wilbur Theater. From our seats on the floor, I craned my neck at the balconies and chandeliers as the lights dimmed and the curtains whisked open to Charlie Brown and Lucy onstage, but my excitement dissipated into disappointment, then guilt, since my father had driven us all the way into Boston, was so excited for us. But Charlie Brown wasn't short and bald with one squiggle of hair and dot eyes. He was an adult man, which was weird. The stage was not a lawn where Lucy yanked the football away but an arrangement of geometric blocks. Soon my disappointment evaporated as the characters, Peppermint Patty, Linus, Snoopy, acted characteristically.

Later, in fourth grade, I was cast as Sally in a Charlie Brown play at school, probably because I was tiny. If I had been cast based on my personality, I'd be Lucy, bossy and controlling and critical. ("We critical people are always being criticized," Lucy says.) Lucy was my least favorite character.

One Saturday we drove into Boston to see the Cowsills at the Hatch Shell beside the Charles River, a free concert that attracted 30,000 people. The Cowsills had seven kids, like our family. The band included the mother, Barbara, with WASPy good looks and *au courant* frosted hair, and five of her six sons — Paul, William, Barry, Robert, and John — handsome boys with cleft chins and mops of hair, and Susan, a year older than me, with a pixie haircut like mine, who made her live-concert debut that day. The Cowsills were the ideal American family, good-looking, clean-cut. In my memory, I linked the Cowsills with milk, the image of scrubbed wholesomeness. I thought this association was because of the *cow* in their name, but I'd forgotten that the Cowsills promoted milk in ads for the American Dairy Association.

For a brief moment my mother dreamed that we, too, could be a family band, like the Lennon Sisters, who were nine, twelve, fourteen, and sixteen when they debuted on *The Lawrence Welk Show* in the mid-1950s. My mother sat us on the piano bench in the living room. "Sing!" she implored. "You could be the Lennon Sisters." She didn't know how to train us, how to teach us to harmonize, but she'd done her part, bearing four girls in four years, so by dint of that feat we should take it from there. "Sing," she beseeched.

In Boston we parked a mile away, it seemed, and held hands as we weaved through the crowd, the paper-doll chain of us traipsing through the city streets to the field, where my mother spread a blanket and unpacked lunch. When the band played, the audience roared. My father lifted us one by one to his shoulders, and from that vaulted view I saw tiny Susan Cowsill way up on the stage tapping her tambourine as the speakers blared the hit song that everyone recognized, the crowd cheering as the Cowsills sang, *flowers in her hair, flowers everywhere / I love the flower girl,* the family band and our family that day, as the song promised, *happy happy happy.*

On Saturdays my father took us to the library or to the park, or my mother took us to the town dump to pick "perfectly good" castoffs, or to Walpole Prison, the Hobby Shop, where inmates sold crafts. We'd park in front of the prison, walk up the main steps, and pass through two tall wooden doors. To the left of the lobby — if that's what you called the entrance; it seemed a little casual, like the lobby of a theater — was the Hobby Shop, crowded with bargain-priced furniture and dusty glass display cases, like waterless aquariums, filled with leather crafts, belts and wallets. On top of the cases were ceramic figures and things made from Popsicle sticks — two-foot-tall Popsicle-stick lamp bases and Popsicle-stick bowls. The heinous prisoners, the worst in the state, murderers and rapists, practiced leatherwork and ceramics, arts and crafts, like we did in school. In spite of my mother's warnings, I thought that prison must be like art class. I couldn't reconcile the image of scary bad men holding tiny paintbrushes and squeezing tubes of glue, licking all those Popsicles to get the sticks.

At the store one day my mother bought a cheap nightstand, a bookshelf, and, for Patrick's room, a Batman lamp, which I coveted. I loved Batman, though I was frustrated by his ineptitude; he was always *almost* defeated, tied to that conveyor belt pulling him ever so slowly toward the buzzing circular saw. Batman's power was punching. *Pow! Bam! Zonk!* To save the city, Batman had only rope for scaling and rappelling, his fists, a fast car, a young friend, Robin, who didn't contribute much, and a butler. Mainly Batman had money. Money was his most powerful superpower. Batman didn't have to work or cook or clean. Alfred the butler did all that and so Batman had time, time to fight crime.

Catwoman was my true hero. She was a criminal, but she couldn't have been *that* bad because Batman had a crush on her. I wanted to live in Gotham City, where girls could be criminals, where the good guy could secretly love the bad girl. Who didn't want to be Catwoman in that skin-tight black bodysuit, sexy, living on the other side of the law, an outlaw?

When my grandmother, my mother's mother, visited from New York, we took her to the prison, a sightseeing stop. In the Hobby Shop one day my grandmother struck up a conversation with the inmate working at the counter. "Why did you do something bad to get in here?" she said. "Why didn't you get a job?" My grandmother — a roly-poly four-foot-ten-inch Italian woman with a sixth-grade education, widowed at thirty-six, who always wore thin cotton print dresses and nylons with clunky black orthopedic shoes for her arthritic feet, who from the age of twelve to seventy-five cleaned other people's houses, scrubbing floors, ironing, washing laundry for people far wealthier than she'd ever be — my grandmother couldn't understand why someone would steal or rob or commit crimes.

The man working in the Hobby Shop that day was most likely Ronald "the Pig" Cassesso, a Boston mobster sent to Walpole for a gangland slaying. For many years Cassesso ran the Hobby Shop, which was outside the prison's secure area; Cassesso even had his own office in the lobby. If Cassesso responded to my grandmother, it's lost to memory, but she couldn't let go of her inquiry. She turned to a woman in the shop. "Why can't they do this instead of going to prison? Isn't it a shame? All that talent," my grandmother said, looking at the furniture, the leather crafts. "Can't they make a living doing this?" A bell rang in the Hobby Shop, and the woman with whom my grandmother had been chatting left to visit her son.

At the Hobby Shop my mother said I could choose something, but the store didn't seem to carry many items for girls. There were fancy dollhouses — the craftsmanship so fine they were sold in FAO Schwarz, too

—but I knew not even to ask for something so expensive. I must not have seen the choker necklaces with earrings to match, handmade by Albert DeSalvo, the Boston Strangler, who was incarcerated in Walpole Prison during the years my mother took us to the Hobby Shop.

DeSalvo was accused of viciously raping and murdering by strangulation thirteen women in eighteen months, though he was convicted on unrelated charges of burglary and rape. DeSalvo confessed to being the Boston Strangler after he'd been caught as the Measuring Man, a rapist who knocked on women's doors posing as a model's agent, telling the women he'd noticed their shapeliness and asking if he could take their measurements. He was also identified as the Green Man, a rapist in green work clothes who knocked on women's doors and said the landlord had sent him to fix the plumbing. The Boston Strangler was the first televised serial-killer case of the twentieth century, so it was international news from the time of the killings in 1962 and 1963 and throughout the three-year manhunt.

In February 1967, when I was seven, DeSalvo was caught and sent to Walpole, bringing infamy to the prison and the town both. Everything the Boston Strangler did was of intrigue, deemed newsworthy: his first breakfast in Walpole Prison (scrambled eggs, toast, bacon, coffee, orange juice); a lawsuit, brought by his famous attorney, F. Lee Bailey, seeking millions from 20th Century Fox for portraying DeSalvo as "vicious" and "depraved" in *The Boston Strangler* (though DeSalvo was pleased that Tony Curtis played him); DeSalvo's bid to be tested for a chromosomal abnormality, which might explain his violence; his dance with an elderly woman, Mrs. Mary Monroe of Jamaica Plain, at a picnic in the prison meant to unite two "neglected" groups, senior citizens and inmates; the 45 rpm DeSalvo recorded in prison, "Strangler in the Night."

I'd heard vague child-whispers about a killer and his sex crimes. The Boston Strangler choked some victims with their own nylons, and though this was not the ligature used to strangle all the women, it became his signature, at least in my mind. I couldn't lose the image of flesh-colored stockings around a neck, silky beige stockings like the ones in my mother's top dresser drawer. There was a confusion in my mind of murder, sex, rape, especially since the handsome, sexy actor Tony Curtis was cast as DeSalvo in the movie.

Whenever I heard about an escaped prisoner, I imagined the Boston Strangler, because he was the only prisoner I'd heard of in Walpole. One summer night Sally and I and some neighbor girls slept out on our screened-in porch. In the early misty morning as we walked around the front yard in our nightgowns, our feet wet with dew, we spied a figure walking down the road toward us, a silhouette emerging through the fog — a man, an escaped prisoner. *The Boston Strangler!* We ran inside the house and grabbed a carving knife from the kitchen drawer, then crept onto the porch. We peeked into the street and *there he was in front of our house!* We screamed and in a panic piled onto each other. When we stood up, the knife was underneath our heap of bodies, the man gone.

That was 1969, the year the Rolling Stones released "Midnight Rambler," based loosely on the Boston Strangler, who, as Mick Jagger sang, would *stick my knife right down your throat, baby, and it hurts.*

Our games were tinged with darkness, the prevailing mood as the 1970s began. Sherry Stewart had her own room and a record player, so we played DOA at her house, inspired by a song with that title. We pulled the shades to dim the room, and I lay on Sherry's bed as she set the needle on the 45 rpm, faint sirens whining through the scratchy speaker. *I remember / we were flying along and hit something in the air.*

Sherry, the nurse, tickled my stomach to mimic blood trickling, but she didn't touch my arms because they'd been blown off. Sherry placed a towel over my legs because they were gone, too. I conflated this song, "D.O.A.," with *Johnny Got His Gun,* a paperback I'd read about a World War I soldier who woke up in a hospital as a quadruple amputee, and with the Vietnam War, ever-present in the background of my childhood. I stared at photos in my father's magazines — a naked girl screaming, people on fire — but in school we didn't talk about Viet-

nam. On our street the fathers were too old to be drafted, their sons
too young. The only person I knew who went to Vietnam was Arnold
Logan, who'd babysat us a few times at our first house in East Walpole.
One day suddenly Arnold Logan was our mailman, walking down the
street in slate-blue shorts, carrying a bulging leather satchel, after our
regular carrier, Mr. Mancini, had been caught with a cache of samples
in his basement, cartons of tiny detergent boxes and mini-toothpastes
meant for postal customers.

One morning Arnold knocked on our door and asked for my mother.
She and Arnold chatted while I hovered next to her. Arnold seemed
nervous, unable to hold my gaze. He seemed frightened. When he left,
I asked my mother what was wrong with Arnold, and she said he'd just
come back from Vietnam. Arnold had all his arms and legs, but Vietnam, I thought, had hurt something inside him.


Sherry Stewart's father called me Bunny, but I didn't connect the nick-
name with the *Playboy* magazines he turned cover-down on the coffee
table, or the game that Sherry invented for our Barbie dolls, Playboy
Bunnies. For this game we left the dolls topless as we walked them
around saying, "Cigars? Cigarettes? Or me?" Then we mashed their
faces into the Ken doll. I hated that game but indulged Sherry so we
could play one of her many board games, like Mystery Date, in which
you turned a doorknob to discover which of five men would take you
on a date. If you got the garbageman, you lost. The blond man was the
date you were supposed to want. Secretly I desired the grunge, the
dark-haired man, the underdog. The men were distinguished by their
attire, which indicated the type of date: picnic, surfing, skiing, a formal,
and a guy wearing work clothes, who I thought was a garbageman but
who could have been any blue-collar worker.

Sherry Stewart and I locked ourselves in her upstairs bathroom
and flipped through her brother's *Penthouse* magazines. Although
we couldn't tell from the close-ups of women's crotches, Sherry and
I agreed that the man put his *thing* in the woman's hole, but we didn't
know what he did once it was in there, how long he kept it there. Sherry

thought he peed in there, but that sounded wrong. Why would he *pee* in there? Sherry wanted to try peeing like a boy, so she tied a string around her pink hairless labia and stood in front of the toilet, but urine sprayed all over the seat and dribbled down her leg.

At Carole Kraus's slumber party in sixth grade, five girls sat in a circle in Carole's screened-in porch and one by one kneeled and pulled down their pajama bottoms. I was secretly thankful that I was not as underdeveloped as Sheila Barton, and astonished at the womanly body of Amy Phelps, her thick dark patch of hair. We told sex stories we'd heard from our older sisters. Eileen Gomes told us that her sister, who thought she was pregnant, stuck a hairbrush up herself to "get rid of it." This was 1971, two years before *Roe v. Wade*.

Eileen had vast knowledge because her older sister was much older than mine, and because she'd already kissed boys. Eileen the Experienced told us that her sister had hickeys on her breasts. Hickeys were bite marks, Eileen explained. I wondered why someone would do that, or want that done to her. Eileen's sister shared with her tenets of femininity. "Once you spread your legs for one man, it doesn't matter how many men you sleep with after that." It sounded profound.

Everyone had a story to tell except me; I felt naive and sheltered, so I invented a story. I said that Brad Peterson, the boy next door, whom I secretly adored, tied Maureen Murphy, one of the sluts everyone talked about, to a tree and felt her up. I don't know where this bondage story derived from, and I didn't make the connection that I'd chosen a slut who shared my name and an older boy on whom I had a crush. Perhaps the bondage relieved "Maureen" of culpability. What could she do? She was tied up, she couldn't escape and so could not truly be a "slut." At eleven-about-to-turn twelve, everything was still in my imagination.

# 2

## *Tilt*

AS THE 1960S SHIFTED INTO THE 1970S, THE DECOR OF OUR HOUSE shifted, too. We knocked down the wall between the kitchen and dining room, crashing a boundary, opening a space. We replaced avocado-green and earth-brown carpets with mauves and cool blues, like the cheap mood rings that everyone wore at school, green dissolving to blue, though after a while the rings turned black and fixed that way, as if to reflect the zeitgeist. In fourth grade, at recess on rainy days my friends and I played 45s, dancing in the back of the classroom to the top hits of 1969.

> *Take a letter, Maria, address it to my wife,*
> *Say I won't be coming home, gonna start a new life*

And "Leaving on a Jet Plane" by Peter, Paul, and Mary, *I don't know when I'll be back again.* A husband who cheated, a wife who cheated, marriages falling apart, people leaving. I was too young to read the auguries in the lyrics, just the solemn tone of Mary Travers's refrain, *I'm so lonesome I could cry.*

Changes, small at first, seeped into my awareness. Off the back porch of our house, a metal garbage can was sunk three feet into the ground, with a step-pedal lid that opened to reveal decomposing slop and wriggling maggots, emptied weekly by an unfortunate man in a golf-cart-like vehicle. Then one year — lo! — trash and garbage were merged and the garbageman became extinct. Then the milkman went the way of the garbageman. No more would anyone find deposited at her doorstep four heavy glass bottles of cold delicious milk sealed with waxy paper caps, set into an insulated aluminum box like an anonymous gift. The garbageman, the milkman — I mark those changes as my first conscious moments of resistance to the future, as if heavy invisible hands on my back were pushing me forward. The future was depicted every Saturday morning on *The Jetsons*. We would eat dinner as a pill, walk on a conveyor belt to nowhere, the treadmill rolling faster and faster, threatening to suck you under. That treadmill clip made me anxious. You had to keep up; the future was coming fast.

Until sixth grade, girls were not allowed to wear pants to school. At Fisher Elementary we stood outside before the bell rang with reddened legs and numbed thighs, waiting in subfreezing temperatures for the doors to open. Then in sixth grade the no-pants rule vanished mysteriously. I didn't know that Sue and Sally had taken part in a protest at West Junior High for the right to wear pants, the seventh-, eighth-, and ninth-grade girls refusing to enter the school until the principal came out to negotiate. They struck a deal allowing girls to wear pants on Fridays, but once they'd crossed the pants line, there was no returning. The girls wore pants whenever they wanted.

At first the changes seemed to originate from beyond our dead-end street, like the DDT trucks that cruised slowly past our house, releasing a fog of poison that drifted into our yard, into our neighborhood, settling here in the dead end. Maybe the bad news had always been there, but I'd reached an age when I saw it, noticing things one year that I could never *not* see anymore. One summer the Gibsons' lawn turned patchy overnight, it seemed, the way fall arrives not on September twenty-first but on that first morning when blades of grass are sheathed in frost and it's clearly the end of one season, the beginning of another. The blight seemed mysterious, though it was likely caused by the weedkiller Mr. Gibson sprayed on crabgrass, a scourge that hit suburban lawns like an epidemic. "The weed has become a neighborhood problem, like juvenile delinquency," *Time* magazine reported.

Then Mrs. Peterson got breast cancer and neighbors sent their children over bearing casseroles. We all knew that Mrs. Peterson had to wait five years to see if the cancer would return, a ticking time bomb inside her. And then the McKenzies' new baby died of crib death. Crib death seemed like an omen — its quiet nighttime work, its tiny innocent victim. I asked my mother *how* the baby died, but she couldn't explain. A blight, cancer, a death — in my memory these events are clustered like beads on the rosary my grandmother worried with her fingers.

Like a contagion, the bad news entered our house just before I turned twelve, on a spring night when my mother called my siblings and me in from playing kickball for a family meeting. The four oldest girls — Susan, Sally, me, and Joanne — lined up on the couch like starlings on a telephone wire. Patrick and Barbie shared the piano bench. Mikey, just two, was upstairs in his crib. My mother and father sat next to each other on the gold ottoman. "As you know," my father began, "your mother and I haven't been getting along." That's where my father's speech ended for me. My mind was puzzling with "as you know." I'd never seen my parents fight, so I had no clue that anything was amiss. "Not getting along" was visible, tactile, aural — the dull thud of my fist on my sister's back, board games overturned, Monopoly money fluttering in the air, the names we shouted at each other. *Fatso. Pimpleface. Twisty-teeth. Bucky beaver.* How could I have missed my parents "not getting along"?

At that moment all the distractions — the shouts of kids playing in the street, the abrasive upholstery of the couch scratching my bare thighs, the bothersome warmth of my sister's arm brushing mine — faded to background. My eyes fixed on my father's lip quivering. I couldn't stop staring; it was so strange, his lip shivering on a warm spring night. My stomach clenched. A tear slipped down my father's cheek, and then like a chorus we all cried, our last act as an intact family.

After the meeting we were sent upstairs to our rooms, though it was still light outside. When one of the neighbor kids knocked to see if we were coming out to play, my mother said, "No, they are staying in for the night." In my room I turned over that word: *separation*. It sounded like something my parents had no choice about, like when I was separated from Martha Wilkins at school for talking during class or passing notes, the teacher moving our seats to opposite sides of the room. This separation felt more ominous, like a rent in the earth, or the planet tilting too far. In fourth grade Mrs. Kannally drummed into our heads the reason we had four seasons, "because the earth is TILT-ed on its axis," she said over and over, emphasizing her point by tipping the globe she held in her hands. That's how it felt, the earth tilted and things falling, tilt like in a pinball game when you slammed the machine too hard, tilt and the game was over. Tilt, the family was over.

What was I doing when my father carried his clothes out of the house and never came home again? I have no memory of him walking out with his arms loaded, stuffing things into his car, only that he was gone overnight, which seemed sudden and devious; there had been a plan all along, and they'd only clued us in at the eleventh hour. Maybe if we'd been warned, we might have tried to convince them otherwise, the way I'd tried to persuade my mother not to give away the cardboard-box dollhouse. Were my parents aware that they'd announced my father's moving out in that bubble of time between May 1, 1972, when President Nixon proclaimed Father's Day a new holiday, and six weeks later in June when it was celebrated for the first time?

My father didn't take much when he left: his tall dresser, on top of which was a tray with cufflinks, his watch with a gold spandex band, the supply of coins replenished each night when he emptied his pockets after work. What I missed most, what seemed starkly absent, were his things in the medicine cabinet in the bathroom: eyedrops and nasal spray for his hay fever, his toothbrush with that weird rubber fin, the minty deodorant, the barbershop-striped can of shaving cream, his razor blades in their little metal container like a suitcase.

Weekend mornings my father faithfully did situps on the floor in his bedroom, wearing a white half-sleeve T-shirt and pajama bottoms. Sometimes I'd sit on his ankles to weigh down his legs as he curled. Afterward I'd sit on the toilet and watch him shave, studying him as he scraped the razor across his throat with quick, sure strokes, the sandpapery sound as he drew the blade along his jawline. "Doesn't that hurt?"

I'd say. "Why don't you grow a mustache? Did you ever have a beard?" He'd wipe his face with a cloth, missing specks of shaving cream. "You left a spot under your nose." He'd splash amber liquid from a clear glass bottle into his hands, then pat his cheeks, his skin so pale it seemed translucent, like wax paper, more so in contrast to his coal-black hair. When he left the bathroom to get dressed, I'd run the water in the sink and push the pepper specks of stubble down the drain with my fingers until they were gone. When my father moved out, he forgot his bottle of orange-colored aftershave in the medicine cabinet, and once I unscrewed the round black cap and conjured him.

The separation was an end to intimacy with my father, an end to seeing him in his pajamas day after day, seeing him as ordinary and vulnerable and human — sleepy, crusty-eyed, unkempt, knowing him in all his moods, watching him in his morning rituals, situps, shaving, coffee, the newspaper or *Time*. I'd hover near him, close enough to smell his coffee-scented breath when I asked for a sip, black and bitter. One of the last images of my father living in our house is this: he is strung up in the doorframe of my parents' bedroom, rigged to some counterweighted contraption that lifted his head and somehow helped the pinched nerve in his shoulder. He sat in a chair in the threshold of their bedroom, one foot in and one foot out, barely able to move.


My parents' separation was the first on our street, the first of my friends' and my siblings' friends' parents; we were cultural pioneers in a new landscape. There wasn't even a television show that depicted divorce until *One Day at a Time* debuted three years later, in 1975. One afternoon as I walked down our street, Mrs. Peterson called to me. "Maureen, do you want to see my little girl?" When I was in the bedroom staring down at her dough-faced toddler, Mrs. Peterson said, "I don't see your father's car in the driveway. Did he move out? Are your mother and father getting a divorce?" I walked fast down the hallway and out of her house.

But we were not alone. My parents' separation coincided with the beginning of what demographers call "the divorce boom." Marriage breakups more than doubled over the decade of the 1970s, peaking in 1979 at about 50 percent and hovering there since. In each year of the 1970s, political commentator David Frum wrote, "one million American children lost their families to divorce." I can see them, fathers with suitcases or garbage bags of clothes, loading up their cars or pickup trucks, backing out of driveways, waving or not; maybe they didn't want to look back as they slowly pulled away, as my father drove up our dead-end street, an exodus of fathers, the breakup of the nuclear family like a chain reaction. It was nobody's fault, since there was no-fault divorce, a law passed first in California, effective January 1, 1970, as if opening a gate into a new decade. The bill was signed by Governor Ronald Reagan, who in 1980 would become the first divorced president.

My father moved into a dank apartment on Savin Avenue in Norwood, the next town over, in a neighborhood of closely situated two-story houses turned into apartments. I'd never been in an apartment before; my reference point was that comic strip in the Sunday paper, *Apartment 3-G*, which I never liked because it wasn't funny. Why was it a *comic* strip? My father's college friend, Dick Walsh, was released from Walpole Prison around the time my father moved out — Dick had been convicted of rape — and a mutual friend asked my father if Dick could stay in the apartment, at least for a while, but my father said no, he couldn't manage it because he was going through a separation.

Perhaps my father was worried about us staying overnight in his apartment. The four big girls — Sue, Sal, Mo, Jo — slept over there two or three times, but then never again because on weekends we wanted to go to dances at Blackburn Hall or sleep over at our friends' houses. My parents wanted us to have normal adolescences, I suppose, didn't want us to resent being stuck home on a Friday night with our father.

What did we do those few nights at my father's apartment? It seems we had to *do* something; we couldn't just be, as we had been at home. Did we play games? Watch television? Instead of our father living with us, he became like a relative, someone to visit, a chore, an obligation.

His apartment was depressing, with sparse furnishings, a cheap secondhand kitchen table, mismatched chairs and lamps, a rollaway cot for a couch, a small black-and-white television set on a fold-up tray table. The place emanated loneliness, containing only my father.

For months after my father moved out, he came over every night after work and he'd play the piano before dinner and it seemed like nothing had changed, even though he left at bedtime. At some point, maybe a year into their trial separation, my parents' marriage counselor told them that my father was coming over too often, that they were not *letting go,* that my parents must *separate.* From then on my father had set visiting hours, Wednesday nights, and Friday nights when he picked up Barbie and Patrick and Mikey to sleep over, though eventually it would be just Mikey. That's when my father began to drift away from me, or I from him.

When my father stood in the kitchen on visiting nights, the burden became to talk, and so the talk became strained. A new formality crept into our relationship, a stiff awkwardness. At first he'd walk into the living room and play the piano as he waited for the younger kids, but as time passed he stopped, as if he needed an invitation into the rest of the house. I could not bear to see my father dumbly standing in the kitchen on Wednesday nights in his slack business suit, tie loosened or undone, his shoulders sunk as if he had no bones, so utterly changed that he was unrecognizable, or on Friday nights when everyone was running around making plans for the evening that didn't include him.

Often I didn't come downstairs to say hello. I pretended I wasn't home, or I left the house before he arrived. If my father tried to hug me, I moved around him like a basketball feint. "I don't care," I'd told my friends when my father left, and bragged that now I could get away with stuff.

That June after my father moved out, near the end of sixth grade, I came home from school one day to find huge boxes in our backyard, a pool kit my mother bought with money she borrowed from her mother and by draining all of our bank accounts: the $10 deposits from birthdays, First Communions, Christmases — she literally pooled our savings. For years we'd clamored for a pool. In summers, stuck in traffic on the way home from the beach in the stifling un-air-conditioned station wagon, with seven sandy, salty, cranky kids, the faint low-tide stench of a smuggled horseshoe crab, my mother would say, "We wouldn't have to go through this if we had a pool." But my father was reluctant: the danger, the expense.

Maybe my mother wanted to compensate for our father's leaving by giving us the one thing we wanted most — a pool. Or maybe she thought a pool would distract us from the separation.

Maybe now that she was on her own for the first time in fifteen years she could make her own decisions, do whatever she wanted.

The kit was for a sixteen-by-thirty-five-foot oval-shaped above-ground pool with a six-foot-deep hopper, which the salesman promised would collapse if installed in-ground, as was my mother's plan. Mr. McGrath, our neighbor across the street who installed pools for a living, warned her, too, but she was undeterred. "Come hell or high water," she always said. My mother hired an excavator, Leroy Jones, a barrel-bellied guy who charged just $30 to dig a hole in our side yard. Leroy felt sorry for my mother, who clearly had no money and who was installing a pool with the help of children. All day Leroy scooped earth with his backhoe, chomping a stump of cigar.

Toward the end of the day, Leroy steered his bucket for another bite of earth, but somehow he miscalculated, and before anyone could yell a warning, the backhoe keeled into the hole. Leroy must have felt the weight shift beneath him, because, tubby as he was, he leapt from the driver's seat before the truck landed on its side in our future pool. Neighbors gathered to watch a tow truck winch the backhoe out of the pit.

Leroy's excavation wasn't quite deep enough, so each day we shoveled dirt and added it to a giant mound Leroy left. For days and days we dug and shoveled and pushed wheelbarrows of dirt into our side yard, spreading it evenly over the ground, digging and shoveling and hauling under the broiling sun like members of a chain gang. Sometimes Drew Peterson from next door helped, but the burden of the work was left to a skeleton crew: my mother; me, twelve; Joanne, eleven; Patrick, nine; and Barbie, eight. Mikey was two. Sue was fourteen and had gotten her working papers, basically a permission slip from a parent, and been hired at the brand-new McDonald's. By dint of birth order as the oldest, or maybe because of her empathic disposition, Sue took care of the rest of us, which included paying for food and pool supplies with her paycheck or bringing home leftovers from McDonald's.

Sally, at thirteen, was inside making lunch or dinner. Sally had little to work with as we were low on funds, but she had a flair for cooking; she faithfully watched TV shows I thought painfully dull, Julia Child and *The Galloping Gourmet*. When we played restaurant as little kids, I was the waiter, with a neatly folded dish towel draped over my arm, and Sally, the chef, prepared hors d'oeuvres, peanut butter toast cut into bite-sized pieces topped with a square of Life cereal. For her tenth birthday Sally asked for homemade ravioli for her special meal, she and my mother all day hand-rolling sheets of pasta, filling and cutting, the kitchen a floury mess.

My role, it seemed, was to complain. One hot miserable day we were digging as usual, using frying pans to scoop soil since we had only two shovels, my mother tanned and muscular, sweating and powdered with dirt, wearing shorts and a bra. She began each morning dressed in a sleeveless tank top, but in the heat of the day she'd decide that a bra was equal to a bathing suit top. "How come Sally gets to make lunch and we have to dig?" I said. "How come Sue gets to leave?" My mother threw a shovel at me, dirt and all, which lightly grazed my thigh. It was more of a *letting go* of the shovel, her grip on the handle relaxed by my lullaby of complaints. I was sent to my room, where I read a book, trying to ignore the *clink-clink* of the frying pans ladling dirt outside my bedroom window.

Because the excavation was too shallow, our pool would be half inground, half above. With the pit cleared, we dug holes two feet deep around the perimeter to sink steel support posts. After we cemented the posts, my mother realized that they'd rust underground, so Barbie, the only one tiny enough to crawl into the holes, was given the job of painting each post with a viscous black sealant. After she finished the first post, Barbie looked up and saw post after post after post — one every three feet all the way around the pool, sixteen total. *Just keep going,* she told herself, *just keep going.* Barbie was the opposite of me — persevering without a grumble, covered head-to-toe in epoxy by the time she was done. "Tarbaby!" we called her.

A plumber installed a pipe from the filter to the drain at the bottom of the hopper, and then my mother ordered two tons of sand, which was dumped in the exact spot where Leroy Jones's mountain of dirt had been; I was not happy to see another pile of dirt to be moved. We had to spread the sand evenly over the bottom of the pool, so for days we shoveled and tamped and leveled. While the rest of us slept off the work of the day, at three in the morning in the cool air my mother kneeled in the pool with her rolling pin, spreading and smoothing the sand like a giant pie crust. I imagine her insomnia was a result of the changes in her life: sleeping alone for the first time in fifteen years; her funds halved; a brood of kids to raise without the full-time help of a husband. But maybe those nights were rare moments of quiet for her, and the rolling and smoothing under the moonlight a soothing meditation.

After we bolted the corrugated aluminum shell to the posts, the last steps were to lay in the liner and fill the pool, but first we had to erect a six-foot fence as required by Massachusetts law. My mother rented a post-hole digger, but her arms were too short to work it effectively, so, using the long-handled frying pan, she and Patrick dug as far into the earth as the length of their arms, then sunk the fenceposts. We couldn't afford expensive stockade, so we bought four sections of spaced-picket fence, like stockade with gapped teeth, a space between each plank, and installed these on the street-facing side. Around the sides and back we jerry-rigged chicken wire to the posts, purchasing lengths of stockade one at a time as we could afford them; it would be years before the pool was properly enclosed with stockade.

The final chore was laying in the liner. Mr. McGrath, the professional pool installer, had promised to help my mother with this tricky task. Occasionally that summer Mr. McGrath stopped by after work to check our progress, smoking a cigarette. I sensed his contempt at my mother's quixotic project, installing a pool herself. *Ha!* On the filling day, all of the kids and Drew Peterson from next door were needed to hold the turquoise vinyl liner that we draped into the perfectly smooth hopper and over the top of the aluminum wall, like laying a pie crust into its pan. My mother sent me over to tell Mr. McGrath we were ready for his help. Sitting in his den watching TV, he said he'd be over after that inning.

We turned on our hose and slowly the water rose in the pool, inching upward as we clutched our sections of liner like a great blue quilt cut from a bolt of sky. My mother orchestrated the filling, directing each of us to yank tight or relax. The tension of the liner increased as the water deepened. Our fingers ached from gripping the vinyl. We waited for Mr. McGrath. My mother stepped delicately inside the pool to pull out wrinkles, trying not to leave heel-print craters in the packed sand underneath. With the pool half full the wrinkles were mostly smooth, except for one thick fold. But that wasn't our biggest problem: the liner was dangerously uneven, hanging a foot over the edge of one side but only an inch on the other. If the liner slipped from our fingers, the water would rush in and ruin the painstakingly smoothed sand. We'd have to redo much of our work. My mother ordered all the kids to the short side to desperately hang on to that inch of vinyl.

Mr. McGrath finally showed up. "The way to remove wrinkles is to reverse your vacuum cleaner and blow them out," he said, but it was too late. With the pool nearly full, there was nothing we could do about that one long wrinkle — the wrinkle my mother stared at, that irked her for years. Mr. McGrath lingered for a minute, pronounced our pool fine, then returned to his ballgame.

"Fat lotta help he was," I said.

My mother wouldn't indulge my sentiment. "I hope you become a critic when you grow up," she said. "You always find something to criticize."

Her comment stung. What I said was true, wasn't it?

Immediately after we filled the pool the water turned chartreuse. We had to shock it with chlorine and wait patiently for three days until it cleared. The chlorine burned our eyes, but it was heaven to jump into the pool we'd spent all summer building. We didn't have a diving board, just a wooden utility spool from the electric company. From that platform I practiced back dives, arms over my head, spine concave, striving for height and depth in one move, hurling myself backward.

I joined the town swim team. The first day I could barely swim one length of Center Pool downtown, but the coach, Dick, gave me tips during the early-morning practices, and by the end of the summer my time was fast enough to compete in the last meet of the season. After the season ended, Dick, a wiry handsome man who wore a Speedo in solidarity but whom I never saw in the water, chose five girls to represent Walpole in a mile race. Each night we practiced, lap after lap after lap, until my belly was full of accidentally swallowed pool water, until I burped chlorine breath. The race was held at Farm Pond, a half-hour from Walpole. Fifty girls aged twelve to eighteen bunched together on the shore, five swimmers from each of ten towns, and when the starting gun shot, we crashed into the lake.

On that windy day in that cold choppy lake, waves slapped my face and flooded my mouth when I raised my head to breathe. I was a crooked swimmer, but in the town pool I tracked along the blue lines painted on the bottom. In our pool at home I used the wrinkle in the liner as a compass needle, but Farm Pond was a greenish murk. Every few strokes I'd surface to sight the sailboat that marked the halfway point and correct my course, swim a few strokes, correct again, zigzagging along. Already three of my teammates had climbed into one of the boats that trailed the bodies bobbing in the water. On the home stretch I put my face in the water and paddled my arms and when my knees dragged on sand I staggered up the shore like a primeval creature evolving to a higher order.

One Saturday I joined a charity swimathon, collecting dimes for laps. Once I warmed up I felt I could swim for miles. I counted one hundred lengths and then lost track as I crawl-stroked down the lane, flip-turned, again and again until I was no longer in my body. I was dreaming and thinking. It was so pleasurable in the isolation tank of the pool that I never wanted to stop. After I passed the point of tiredness I felt only fluid, as if I were some sleek creature gliding through the water. If I kept swimming I'd develop gills on my neck like the Incredible Mr. Limpet and be able to live underwater, live in a giant pink nautilus like in my favorite book in third grade, *Dr. Dolittle*.

At the end of a lap as I was about to flip-turn, a hard tap on my head interrupted my reverie. It felt strange to surface, to remember where I was and why. A lady from the event crouched down and said, "I think you've done enough." Her tone implied that I was greedy, swimming excessively, even though I was raising money for charity, even though I felt as if I could swim for hours, for a week straight, swim into the next phase of my life, into adulthood, like Cheever's character in "The Swimmer," who swam home by dipping into his neighbors' pools, swimming the length of one, hopping out and then into the next pool, looking back on his life, crawl-stroking through his past, but instead I wanted to crawl-stroke forward, swim through water like time, swim into my future.

# 3

## *Operation Pocketbook*

ONE DAY AT FERNANDES GROCERY STORE, WHERE EVERYBODY knew us, I watched my mother tuck a can of tuna fish in her pocketbook, the kind we liked, Geisha white tuna in water, not that oily rank cheaper brown tuna. In front of the refrigerator case, she slipped a package of ham in her purse. I glanced around anxiously, but nobody noticed. Who would suspect a tiny housewife-looking woman with a pretty face and rollers in her hair, wearing sneakers with Peds?

Once my father moved out and he had to pay for two households, we had no money — my father paid child support but no alimony. My mother must have felt desperate at our whining about nothing to eat. One night she cooked a bag of frozen ravioli and when the bowl got to Sue, there was only one left, so my mother ordered each of us to pass a ravioli to Sue, a reapportionment. We weren't poor like Doreen Randall, the standard-bearer of poverty in my youth, or the Wests on the skinny street downtown, but my friends weren't eating farina for dinner. My mother tried to pass off Carnation instant milk, but we complained loudly. We didn't understand the economics of divorce.

What clicked in my mother's mind the moment she crossed the line from who she'd been — a good Catholic, teaching Sunday school, working so hard on that dollhouse to raise money for the church, serving fish on Fridays, taking us to confession on Saturdays, Mass on Sundays — to break a commandment: Thou shall not steal? Years earlier, when Joanne was five, she stole candy from Woolworth's one day, so my mother dragged all of us back to the store to witness a sniveling Joanne apologize to the cashier, her pudgy hands reaching up to place money on the conveyor belt.

Maybe Joanne had been confused by my mother's rules or the sometimes ambiguous clauses. Before at Fernandes, if we found a bag of candy on the shelf that was *already open,* it was fine to take a piece. You must never open the bag yourself; that was stealing. The former was acceptable because the opened bag would have to be thrown away and it would actually be a waste *not* to eat that candy, which jibed with the lesson that we should never waste food because children were starving in Biafra. In the candy aisle of Fernandes, we'd search and search for open bags, which we almost always found, some reliable thief preceding us.

When I suspected my mother was going to steal something—she'd glance around casually, her hand surreptitiously opening her pocketbook—I'd walk away, pretend to be looking at something farther up the aisle. If I blinkered myself to the criminal act, then I was neither complicit nor witness. But I grew increasingly nervous when she shoplifted, in inverse proportion to her skill; she got good at it. Once, she slid windshield wipers down the leg of her pants and walked out of the store. Still, I worried she'd be caught red-handed and there would be a scene in the grocery store, not the scene we used to create before the separation, when my father was paid once a month and at Fernandes we'd push two carts heaped with packages, the line accruing behind us at the checkout, then fill the back of our station wagon with a dozen bags, a literal mother lode.

I refused to stand in the checkout line when my mother paid with food stamps. The ritual was attention-drawing and humiliating, separating the disqualified nonfood items—toilet paper, soap, toothpaste (as if those weren't necessary)—to pay for with cash, tearing each perforated, colored Monopoly-money bill from the booklet as the line of shoppers behind us grew impatient and, I suspected, judgmental: potato chips with food stamps?

Our fall from the middle class to welfare coincided with a plummet in the economy — stagflation, a brand-new term: steeply rising costs, high unemployment, slow growth. In the months after my father moved out, the cost of meat, poultry, and fish nearly doubled. Millions of Americans protested the spike in prices, marching in picket lines as part of Operation Pocketbook and Housewives Expect Lower Prices (HELP). Food prices rose and rose as inflation hit 9.2 percent in the first half of 1973 and President Nixon ordered a sixty-day freeze on meat prices. By the end of 1974 the nation had suffered the worst economic downturn since the Great Depression.

The 1970s turned a lot of otherwise honest people into thieves. One day at Fernandes, where Sue worked as a cashier, a man filled his grocery cart with meat — roasts and legs of lamb and steaks, the biggest, fattest hunks of meat from the store's refrigerated bins — and pushed the cart out the automatic door without paying, loading the meat in his car and speeding off before the cops arrived. The meat thief.

Shoplifting nearly became a national pastime, increasing 221 percent from 1960 to 1973 and then rising 20 percent annually until 1980, when stores installed electromagnetic tags and surveillance cameras. In 1975 shoplifting was the fastest-growing larceny in the country, according to the FBI. "It used to be poor people," a Boston detective said in the 1970s. "Now we get doctors, lawyers, teachers, nuns, priests, ministers, rabbis, you name it." The president of the Massachusetts Retailing Institute called shoplifting a sign of "a very sick society." Everybody was stealing, it seemed, taking what they wanted, needed, or felt they deserved, a kind of delirious slow-motion looting.

If my mother needed clothes for herself, she shoplifted them — her own personal Operation Pocketbook. In the changing room she'd roll something into her purse. Building 19, a warehouse store for liquidation and fire-damaged goods (you could smell smoke on some of the clothes), with sister stores called Building 19½ and Building 19¾, was an easy mark. Dana Fontaine, the second divorcée on our street and my mother's new friend, worked at Holt's, a fancy clothing shop downtown. When my mother checked out, Mrs. Fontaine let a couple of items slip by without charging for them.

In her thievery my mother was a modern-day Robin Hood, like the Wild Colonial Boy. She stole dungarees for Sue's boyfriend, Jeff, and Sally's boyfriend, Kevin, both from divorced families, too, because they needed clothes. Does the functional imperative mitigate the criminal intent? When my mother finally found a job, she earned $4 an hour as a clerk in medical records at Norwood Hospital. She often borrowed money from Sue, from her job at McDonald's, and later, when Sue worked at Fernandes, she'd pass a few items by as she rang up my mother's groceries.

One day at Mimi's Variety downtown, I pushed a Popsicle down the front of my hip-hugger corduroys, my stomach buzzing with nervousness as I walked out the door, the Popsicle lodged against my pubic bone, burning cold against my hot skin. My shoplifting began as a flirtation with danger but became a necessity, as it was for my mother. Child support and food stamps barely covered groceries and toiletries, gas and auto repair, other sundries; there was no money for clothes. Perhaps my mother was too proud to ask my father for more money, couldn't stomach the indignity of asking, which felt like begging. Judging by my father's shabby apartment, I'm sure he was broke.

When I needed money for school clothes, my mother told me to ask my father. I remember the night I asked — I can picture exactly where my father and I stood in the dining room, next to the table where we used to have family dinners, where my mother sewed the jailbird costumes, where she built that cardboard-box dollhouse — the moment distilled because it was the last time I asked my father for money. "How much do you need?" he said. My father, like my mother, had grown up poor, on the middle floor of a triple-decker cold-water flat, where he slept in a double bed with his two younger brothers until he was eighteen. Even as a kid I sensed his anxiety about money.

My father and I stood awkwardly in the dining room, his shoulders slumped as if he carried the weight of the world, his curly black hair grown out for the times, with longish sideburns, and always a five o'clock shadow. He reached in his pocket for his wallet while he waited for me to say how much I needed. I had no idea what was a legitimate amount to request. "Enough for a pair of pants, I guess." My father grimaced as if it were physically painful to pry apart the folds of his worn black wallet. My siblings and I called this expression "the Dad face," which conveyed not disapproval so much as a queasy discomfort. He handed me a twenty. "Is that enough?" I couldn't look in his eyes, downturned at the corners like mine, which lent a serious, even melancholy note to our faces. "Yeah," I said, "thanks," though I knew $20 would buy a single pair of pants, no shirts or underwear or socks or shoes.

I felt awkward asking for money, guilty almost, as if I owed him something in return; my father must have felt uncomfortable, too, our relationship sullied by financial transaction. I never again wanted to be in the position of having to beg. It was easier to steal. In the dressing room of the Levi's store at the mall, I rolled up pants and stuffed them in an empty shoebox in a bag. I stuffed clothes up the sleeves of my giant snorkel coat — those dark-blue polyester military parkas with orange lining and fur-trimmed hoods, suddenly the coat of choice in the mid-1970s. Shoplifting didn't feel like a crime; there was no breaking, no illegal entering. You walked into the store — they *invited* you inside. Stealing merchandise from a generic department store seemed harmless. Taking things felt more like *helping myself* — in all facets of that term.

Shoplifting made me nervous, which sometimes manifested as uncontrollable silent laughing. Once at a somewhat upscale store called Foxmoor Casuals, I watched my friend Paula Fournier shove an entire suede suit — a skirt and matching jacket — under her snorkel coat, and it seemed so preposterously bold that we could not stop laughing, though no noise came out of our mouths. I laughed so hard I had to cross my legs so I wouldn't pee my pants, bent over in the middle of Foxmoor Casuals, tears streaming down my face as I stuffed a blouse down the front of my coat, feeling slightly dizzy and removed, as if I were in a slow-motion film looking down upon myself, or underwater.


On the first day of school at West Junior High, my old friends from sixth grade didn't save a place for me at lunch, so I sat alone until Grace Giordano and Marlene Hample, who were on the town drill team with my sister Sue, sat with me, and thereafter they became my friends. That's how it seemed to work when I was a kid — sudden tectonic changes, after which you were on the other side of a divide. Martha Wilkins, my former best friend, said the next day that she "hated my guts," my stomach and intestines and liver and kidneys, as if I were the Visible Woman, that anatomy kit Sally had, a foot-tall clear polymer shell in the shape of a woman that you filled with red and yellow die-cast plastic organs. Once assembled, you could see her guts, her tiny pink heart.

Marlene Hample had gone to the Barbizon School of Modeling, which advertised frequently on the UHF stations that carried reruns of *Gilligan's Island* and *The Brady Bunch,* and in magazines. "Be a Model. (or . . . just look like one)," read the ad, with a clip-out form to send for information. All the girls fantasized about going to Barbizon, but it was expensive and models were tall, so that excluded everyone but Marlene. We thought Barbizon screened applicants based on beauty rather than accepting anyone who could pay.

Marlene had a crush on Gordon McGee, a Longview Farm boy. Longview Farm, or "the Farm," as we called it, was a branch of the New England Home for Little Wanderers, a residential program for boys with emotional, learning, and/or behavioral disabilities. The Farm was set on 166 acres less than two miles from Norfolk and Walpole Prisons. The geographic proximity of Walpole Prison, Norfolk Prison, and the Farm created a sort of triangle of troubled men and boys — it implied a relationship: starting at the Farm as a delinquent, a boy might wind up as an adult in medium-security Norfolk, continue his life of crime, and graduate to Walpole, the Farm like a training ground, a minor league team, where one apprenticed for a life of crime.

Or not. Sometimes the sequence was reversed; inmates were moved from Walpole to Norfolk after a period of good behavior and then released into society. The Farm's mission, at least, was rehabilitation, education. Every year boys from the Farm were enrolled in Walpole schools. At West Junior High, we could tell the Farm boys: they were tougher, citified, cool. They had histories of trouble, broken families, infractions with the law. They had pencil-thin mustaches and shadows of sideburns, were swarthy and long-haired. They were like young outlaws, and we girls liked them best.

One day at lunch Marlene said she was going to seduce Gordon McGee that weekend. I confessed that I didn't know what *seduce* meant. Marlene and Grace looked at me incredulously, these girls who'd recently taught me how to smoke cigarettes. To seduce, Grace Giordano said, was to *make* a boy like you. I thought this required some kind of magic power, something Marlene had since she'd gone to the Barbizon School of Modeling. Grace and Marlene talked about sixty-nine as a verb. As explanation, Grace drew a picture on a napkin.

In seventh, eighth, and ninth grades I would be schooled by books I lifted from my mother's nightstand — *The Happy Hooker,* by Xaviera Hollander (the name Xaviera itself seemed X-rated, sexy), and *Valley of the Dolls,* by Jacqueline Susann. I skimmed Henry Miller's *Tropic of Cancer,* looking for the juicy parts (declared "non-obscene" in 1964 by the U.S. Supreme Court), and *Everything You Wanted to Know About Sex But Were Afraid to Ask,* with its handy Q&A format. Most helpful was *The Sensuous Woman* (#3 on the *New York Times* bestseller list), which contained such risqué material that its author was known only as "J." There in black and white were instructions for sex, as if you were making a cake, or learning yoga as my mother had in our living room, listening to Jack LaLanne on a 33 rpm, following step-by-step instructions. I tried to grasp the techniques for fellatio: the butterfly flick, the Hoover, the silken swirl, a name that brought to mind my favorite dessert, Whip 'n Chill, a box-mix chocolate mousse that was fluffy and smooth.

One Friday night at a junior high dance I sat next to Ricky Strick-

land on the wooden floor of Blackburn Hall, a small function hall behind the police station downtown. My back pressed to the wall, his
arm awkwardly around my shoulder, a local band hammering "Johnny
B. Goode" or "Jumpin' Jack Flash," I waited for him to work up the
nerve to kiss me. Ricky Strickland had shoulder-length ash-blond hair,
brown eyes, and a mischievous smile. He and Vinnie Fontana, who
had lopsided Brillo-pad hair and a crooked nose, zoomed through the
school corridors as if they were on motorcycles, their hands torquing
the invisible throttles, spraying saliva as they *brrrm-brrrm*ed, race-
walking down the halls. Ricky, this thirteen-year-old boy who occu-
pied the pages of my diary for a couple of months, who was shy about
kissing, who my diary notes would ask me out again and again after
I broke up with him, would land in Walpole Prison before he turned
thirty, for rape, or, in legal terms, "indecent assault and battery on a
person aged fourteen or older."

On another Friday night at Blackburn Hall I left the dance early
with Al Malone, and with three other couples we lay inside the dugout
of a nearby ball field, lined up like sardines. It was strangely quiet, our
desire enormous and silent. Al's kisses tasted like the Juicy Fruit gum
he snapped, but everything about him was annoying. Al gave me a gold
band, though we'd only been going out for nine days. I didn't like the
band, which was ugly and matronly and obviously stolen from either a
department store or his grandmother's jewelry box. I felt stupid wearing the wedding-band-like ring, so I broke up with him. He wrote me a
note saying that he was going to bury the ring in his backyard, a symbol
of grief and loss, but two weeks later I saw the same ring on the finger
of Grace Giordano, shiny and polished. It was a happy ending, though,
because Grace *liked* Al Malone and she *liked* the ring and they stayed
together for a long time — many weeks.

In my diary I'd written that the ring was "gold and beautiful," even
though I hated that horrid ugly ring. But all the girls made such a fuss,
and everyone seemed in awe of Al's gesture, and he, I suspected, felt
magnanimous in his big display of affection, so I went along, pretended
I liked the ring, tried to convince myself by writing it in the diary. This

single lie writ small was a step down a treacherous path of fitting in, being popular, the cost of which was losing something I didn't realize I possessed: myself.

My mother prepared for our teenhood architecturally. She built a room, or at least finished one, in the cellar, a room of our own, a room to contain us, a door with a lock for privacy, to seal ourselves inside, like house arrest. At night, reading instructions from a library book, she cut two-by-fours with a handsaw and framed in walls. She stuffed pink fiberglass insulation between the studs, then nailed sheets of plywood to the frame. Sometimes I helped by holding the wood steady while she sawed, or by hammering finish nails into the plywood. She installed a drop ceiling, a grid of lightweight pressed-paper squares that interlocked, one or another of us handing panels to her as she stood on a ladder. At Building 19 she bought a roll of burnt-orange indoor-outdoor carpet and laid this directly over the cement floor, and thus we christened the Orange Room. My mother obtained a black couch — probably from Building 19 again — and a hanging lamp with ponytailed go-go girls painted on the frosted glass shade. We bought black-light posters, a lava lamp, a beanbag chair. One year for Christmas the four older girls shared a present — a stereo system.

The Orange Room was like a cave — it had no windows, so you could wake up at 3 p.m. and feel like it was early morning, or any time in the timeless room. But the Orange Room hastened time. When I was ten, my parents forbade me to play their *Hair* soundtrack album, because of the song called "Sodomy": *Sodomy / fellatio / cunnilingus / pederasty / Father, why do these words sound so nasty? / Masturbation / Can be fun / Join the holy orgy / Kama Sutra / Everyone*. They'd seen *Hair* in Boston in spite of, or perhaps because of, protests over its "lewd and lascivious" content. The debut performance of *Hair* at the Wilbur Theater was shut down, but eventually the show reopened after the U.S. Supreme Court ruled in favor of free expression. My parents brought home the playbill and the album, but like the court they censored the music. Not long after, in the Orange Room, I listened to Sally's Ten Years After album, the song "Good Morning Little Schoolgirl," with its screeching refrain, *I want to ball you.*

In the Orange Room with Glen Arpin, my boyfriend in ninth grade, I practiced mouth-to-mouth resuscitation, which I'd learned in a junior lifesaving class, and he showed me wrestling moves, takedowns that left me pinned. We progressed from kissing to second base — his hands under my shirt and then my bra — but everything got dicey at third base, the line between acceptable petting and slut territory. Baseball was All-American, as was the sporting event of trying to score with girls — first, second, third base, home run. The body of the girl was the field on which the boy played to win. But how did the girl win? By stopping play? By scoring? We wanted to play, but the rules were unclear, or stacked against us. We had no bat; we had no balls.

In 1971, when I was in sixth grade, Nixon declared drug abuse to be public enemy number one. At Fisher Elementary, I'd signed a pledge to be mailed to me in the future — a vow never to smoke cigarettes, drink, or take drugs. I'd heard about drugs vaguely from songs, like "Lucy in the Sky with Diamonds," a code for LSD, and Grace Slick promising that *one pill makes you smaller / and one pill makes you tall,* Grace Slick singing about the call, calling Alice, like Circe calling us from the past, the 1960s, into the future. The next year, 1972, Nixon escalated his "total war" on drugs and I smoked my first cigarette, drank alcohol, smoked marijuana in a little pipe. The following year, 1973, Nixon declared an "all-out, global war on the drug menace" and formed a superagency, the DEA. Now in eighth grade, I smoked dope regularly and drank nearly every weekend.

One weekend night six girls slept over at Alison Preston's house, our sleeping bags spread on the floor in the partly finished basement rec room, staying up past midnight choreographing dance moves to Crosby, Stills, Nash, and Young's "Chicago," harmonizing as we stepped and turned, kicked and pivoted, singing about changing the world, rearranging it, about rebellion, a harbinger of the path that Alison and I, among these girls, would soon take — *rules and regulations, who needs them?*

One day a couple of months later in that basement, at the built-in wet bar, Alison poured two cups of grape soda and then added whiskey, brazenly, as if toying with the possibility of getting caught, since her father was just three feet away at his desk, working at home that day, his back to us. We drank our plastic cups of booze as we walked back to West Junior High to watch a game after school, the drink hitting me on the mile-long walk. I felt silly and giddy. I said whatever came to my mind, and Alison and I laughed at everything. I felt wild and free, and in that spot in the woods off school property where kids smoked, I swung around the skinny supple trunk of a birch tree. I was not nervous or self-conscious. I was funny, a better, improved me.

Alison Preston was beautiful in a hard way, her alabaster skin poreless, her jet-black hair falling silkily in her face when she bent her head. Her eyes were ice blue, with short dark eyelashes that looked like liner. She couldn't wear watches, she told us. They stopped ticking when strapped to her wrist, as if she had a certain kind of energy unlike the ordinary rest of us, an electric force field radiating from within her, a member of a rare subset of humans, like geniuses, schizophrenics, lefties. Alison was the girl everyone turned to heliotropically when she entered a room. Her hourglass figure drew boys' attention, men's, always, everywhere. Mr. Braun, the science teacher, chose Alison to fetch something from the supply room, then followed her in there and tried to "feel her up," she told us. Was this true? Why else did we consider him a creepy perv? What else could he have done to inspire Alison's exquisite revenge? When Mr. Braun left the room, Alison spat in the Mason jar of distilled water on his desk, and we waited in gleeful horror for him to sip.

Alison was charismatic; it was hard to say no to her. I wrote in my diary, "Alison wants those guys to sleep over Saturday night." When I read this later, at first I thought Alison wanted "those guys" to sleep over at *her* house, but then I saw that Alison had wanted everyone to sleep over at *my* house. "I hope they're not so drunk that they puke because Mom will probably be home. I hope she goes out." Filled with anxiety, against my own judgment, I acquiesced to Alison's suggestion.

Paula Fournier lived a few blocks from me, and so in junior high we walked to school together. Paula was tall and big-boned, with a mane of auburn hair and light-brown eyes to match, more handsome than pretty, the dimple in her chin a reply to her prominent widow's peak. She was quiet, but quick to laugh and game for anything. One day, with Loretta Petty, we shot beers on the way to school. To "shoot" a beer, you punched a hole in the base of the can with an opener and affixed your mouth to this hole. Then you popped the tab on the top and the beer geysered down your throat. The first time I did this — drank a beer in under ten seconds at 7:30 in the morning — I felt a roiling in my empty, breakfastless stomach. Within seconds the beer projected out of my mouth, just as foamy and effervescent as it had been going down. My sensible body rejected the beer. My nonsensible mind paid no attention and tried again.

Somehow we discovered Tango, a premixed screwdriver, vodka and an uncarbonated orange drink like Tang, the beverage of astronauts. Tango became the drink of choice on weekends, when a mob of kids hung around the junior high. Sometimes Paula's brother, Duane, who was five years older, bought for us, or else we stood behind the liquor store and asked a likely-looking stranger, an older boy or man usually. It was amazing how many people were willing to buy alcohol for fourteen-year-olds. Our success rate was 100 percent.

Usually I was in control when we partied, not sloppy like Tina Baronski, who lost a shoe and cried all night searching for it, or Grace Giordano, who peed all over her pants. But one weekend only two of us instead of the usual three girls split a quart of Tango. I took long gulps of the bottle, greedily it seems, and soon I was drunker than I'd ever been. The world around me swirled and there was only sensation — the astringent orange taste in my mouth, the black starry sky, the rich smell of decaying leaves that I fell onto with Glen Arpin, who was kissing me. I felt cool air float over my stomach as he lifted my shirt, his warm hands on my skin, the damp cold ground beneath me, his crotch pressed against my pelvis, his heavy legs pinning me, his hot breath and soft, soft lips.

Then he told me to be quiet — I wasn't aware I was making noise. We lay still, bright lights shining on us, headlights of a police car. The police usually peeled into West Junior High at least once on weekend nights, and we ran from them as if it were a game, thrilling and breathtaking like Flashlight Tag, only this time the cops were "it." The cops never caught us. We knew the woods and we were young and fast. But that night I was too drunk to run. Glen helped me up and we stood in front of the cruiser, squinting into headlights. The cop said, "What would your parents think if they saw you now?" I was a wiseass, a smart aleck, a back-talker. Fresh. "They don't care," I said. It felt like the truth.

Even before my father moved out, there was always the chance of being forgotten, like in sixth grade, when I'd stay late for basketball or softball practice and afterward stand outside the front doors of Fisher School alone, my teammates picked up already, the sky dimming as the janitor passed back and forth with his broad dry mop, glancing at me still there, the chains clanging as he looped them through the door handles, and then my mother would screech into the circular drive. Always after I opened the car door to her apologies I burst into tears. I hated being forgotten, but now I used it to my advantage.

Glen promised the cops he'd take me straight home. He said I didn't live too far. Did Glen walk me all the way home? I lived only a mile away, a fifteen-minute walk through the woods. How *did* I get home? I don't remember, but my diary does.

April 11, 1974 — Went drinking. We were on our asses. The pigs came up and talked to me and Glen. When I got over Nora's house (everyone slept over there) those guys had to undress me and put me in my pajamas because I was dead asleep. Thank god we didn't get caught.

I didn't remember much from that night, but Glen did, and he told his friends that he made it to third base with me, that I was wild, moaning with pleasure, unlike his previous girlfriend, Suzanne, who just lay there "like a cold fish," he reported. He told his chums, who told their girlfriends, who told everyone else, it seemed. On Monday at school, I was the subject of rumors. I was the loose one. I was the slut. I denied the rumors, but several people had seen me walk out of the woods with my belt undone. "I forgot to buckle my belt after I peed," I told Loretta Petty, who I knew would disseminate this alternative narrative, this spin. Walking home from school, Loretta told me what Glen had told everyone I'd said: "Fuck me." Glen said he *could* have, but he'd restrained himself. He was good and prudent; I was bad and easy.

When I heard those words I had a shock of memory like lightning illuminating a dark path, the scene flash-lit, and I knew suddenly that I'd said them. I denied it, horrified that my libido had been unbelted, exposed to the world. I was mortified to be confronted with this outsized desire, embarrassed, but beneath the surface I was secretly *glad* that I was not a cold fish who just lay there. I was confused; I was not a slut, I was a slut. Glen was not a slut. There was no word to characterize his actions. Sluthood was a sorority to which no respectable girl wanted to belong, though every girl was measured by this calculus.

April 28, 1974 — Today we got a new girl, Sarah Macomb. She is a whore, I think. She wore really tight pants and no bra. Someone said she was pregnant but that could be a rumor. She seemed nice.

The summer between ninth grade and tenth was like crossing a bridge. Part of my life was ordinary — riding my bike to Bird Park to play tennis with my friends; babysitting; selling hot dogs in the concession stand at Schaefer Stadium, home of the New England Patriots, where rock concerts were scheduled on weekends; swimming on the town team — while drifting into another world, hearing the call. That summer I wrote about playing Flashlight Tag, but this, too: "Had beers and whiskey this morning with Sally, got pretty drunk. Went to Dairy Bar to get an ice cream. Slept in the cemetery for a while." In June, I wrote, "Me, Alison, and Paula are buying some hits of mescaline. I've never done that before but I'm not scared." Telling myself I wasn't scared reveals to me that I was, that I needed to convince myself otherwise.

One night that summer I slept in a tent in Loretta Petty's backyard with Alison, Paula, and a few other girls. We stayed up late talking, smoking pot, drinking. At four in the morning we took off our shirts and strolled bare-breasted through the empty streets of the projects, half streaking, naked on top, clothed on the bottom, like mermaids, creatures of two worlds, belonging wholly to neither, shapeshifting in that interstice between dark and light, girlhood and womanhood.

The transformation that summer was not physical—I reached my full height of five-foot-two in seventh grade—but something was happening inside me, some too-fast metamorphosis, like the kid in the Wonder Bread commercial who grew before my eyes. Inside I felt pressure, something wanting to burst out that I had to suppress. Some days just being in the world made me edgy and raw, as if my skin were inside out.

> May 26, 1975—I was in the weirdest mood today. One minute I just felt like crying but I don't know why and the next minute I was laughing. We had an algebra quiz and I couldn't concentrate so I got a 44.

My mother noticed that something was wrong with me. She subscribed to *Psychology Today* and took evening classes in psychology at a nearby junior college.

> May 30, 1975—It was such a terrible day. Everyone was getting on my nerves and cutting me down so I started crying at the dinner table. Mom wants me to go to a phycologist [*sic*].

Psychology was my mother's answer to what ailed me, and the answer to what ailed a lot of people then. A survey on college campuses in 1975 found among students a "prevailing sadness." The 1970s were the tail end of psychology's dark ages, madhouses and loony bins — the last frontal lobotomies were performed in the early 1970s — but also the dawn of the self-help age. There was confusion at the crossroads. When it was revealed in 1972 that Thomas Eagleton, presidential candidate George McGovern's running mate, had undergone electroshock therapy, Eagleton was branded too crazy for office. But in 1974, First Lady Betty Ford unapologetically admitted that she regularly saw a psychologist and *Time* magazine named her "woman of the year" for her candor.

> July 14, 1975 — Just talked to Mom. I'm going to see a phyciatrist [*sic*].
> I don't know what I'm going to tell him, but I just want to get myself analyzed. Maybe I have some kind of complex. I have one big, huge cry stored up inside me. It will all come out sometime though.

I wonder what kind of "complex" I thought I had, or what I even knew about "complexes." Maybe I'd found the word in one of my mother's *Psychology Today* magazines. I remember vividly an article about self-perception, which was accompanied by three illustrations of a woman. The first image showed how the woman viewed herself (homely, wrinkled); the second illustration was how the woman thought others viewed her (glamorous, like Ginger from *Gilligan's Island,* copper-haired, with thick eyeliner, a voluptuous mouth); and the third was how she actually looked. The first two images depicted the perspective of someone with an inferiority complex (ugly) and someone with a superiority complex (beautiful), which was "narcissistic." After reading the article, I suspected that I had a superiority *and* an inferiority complex. The article bothered me immensely. What if my perception was skewed?

One day in third grade I came home from school to find that my mother had sketched a portrait of me from my school photo. I was amazed at the resemblance, the lopsided curls formed by pink foam curlers worn to bed the night before, the home-sewn tulle shirt and gray plaid vest, that year's Easter suit. In the photo I showed my big happy smile with the torqued-out teeth. "How did you learn how to draw?" I asked my mother. "I just studied the picture," she said. I was dazzled by her hidden talent, and pleased she'd chosen to draw me of all her kids, but then I said, "Why did you have to draw the crooked teeth?" My mother grew irritated. "Because that's how you look!" She wanted appreciation, not criticism, but criticism was my métier. I wanted my mother to erase the diastema "big enough to hold three pieces of corn," as a boy once said, some random boy standing behind me at the movies. I wanted her to fix — with the malleable lead of pencil — my "twisty teeth." I wanted her to draw *not* me.

July 21, 1975 — Mom, Sally, and Sue are going to New York this weekend. I wanted to go but I could tell they didn't want me. I feel so unwanted lately. I feel like killing myself even. What a sucky life.

Sometimes at home in my bedroom with noise all around me, kids fighting, someone talking on the phone, my mother cooking or washing dishes, clattering pots and pans, "Chopsticks" or a duet banged out on the piano, the radio playing, a television on somewhere, I felt lonely. As a girl I used to love lying on the grass looking up at the sky, thinking and wondering, finding pleasure in the company of my thoughts. But I'd lost that world unto myself. Sometimes the loneliness felt like someone was tapping me on the head, patting me like grownups used to do when I was a child, except now that hand was pushing me down.

July 22, 1975 — Lately I feel so depressed or something. I can't figure out what it is. I need someone to talk to.

I'd agreed to my mother's suggestion to see a psychologist only if he was in a different town. I was so self-conscious that I thought the entire citizenry of Walpole would somehow find out if I saw a shrink in our town — a measure of my small view of the world, or my overblown view of myself. *Narcissist.*

My mother and I climbed a dimly lit stairway in a two-story clap-board house converted to offices, looking warily at each other, the stained walls, the creaky stairs. I don't know how my mother found this guy; maybe she'd looked in the yellow pages. She knocked on the door and the doctor invited her in for a private chat, then she waited in that shadowy hallway for nearly an hour. Dr. Rosenwald was a heavy man with rosacea on his face, patchy red skin like my father sometimes had. The office seemed shabby, with old-fashioned radiators, the paint chipped, worn indoor-outdoor carpet like we had in the Orange Room. The shrink must have been inexpensive, the only shrink my mother could afford, a bargain-basement shrink. That was my impression.

After Dr. Rosenwald asked me a few questions (if I smoked dope, cigarettes, if I drank, to which I said yes), he stared at me while I sat silently for the next half-hour. I wondered why he wasn't saying any-thing, or why he wasn't giving me some type of test — Rorschach or word association or something. After the session was over, my mother took me out for ice cream, a rare time alone with her. Maybe that's all I needed, to go out for ice cream with my mother. I told her I didn't like Dr. Rosenwald, that he didn't *say* anything, but I agreed to try one more time. My second visit yielded only a cryptic note:

August 7, 1975 — Went to the phyciatrist [*sic*] again tonight. He doesn't do any good. That was my last time seeing him.

A week later my father took us on vacation to a cottage in New Hampshire, honoring our family tradition, though it would be the last time. My mother prepared meals for the week so nobody had to cook, her presence felt in spite of her absence. During the day we swam or played tennis, at night played cards or board games. I took my precious diary with me, which shows that my confusion and sadness remained and that I still hadn't learned how to spell *psychiatrist*.

August 15, 1975 — Everyone is getting on everyone else's ass and I'm getting so sick of it. I'm really confused. I feel like crying but I don't know why. I think I have a problem but I don't know what. This happens to me all the time and I never know why I feel like crying. The fucking phyciatrist [*sic*] didn't do a damn thing.

Just after I started tenth grade at the high school, where Sally was a junior and Sue a senior, the "big cry" that I'd felt inside me sometimes seeped out, like a broken water main.

September 6, 1975 — I went downtown today. I was walking alone and I started crying really hard. I don't know why.

I fought with my mother every morning before school, storming out of the house in tears, tinny clatter from the screen door slamming behind me. My mother complained about the way I dressed — "You're not going out looking like that!" — the ratty jeans, the wrinkled, stained dungaree jacket. Every day I rubber-banded my wet hair into a ponytail until it dried to straighten the curls, though it left a weird indent in my hair. My shaggy bell-bottoms dragged on the ground, the hems frayed and dirty, my midriff shirt revealing a patch of skin between my hips and belly button. What were my mother and I really fighting about? Probably she hoped that if that grungy outer layer were stripped, the daughter she used to know would be there. She must have thought that the girl I'd been a few months earlier — captain of the cheerleaders, honor roll student, on the basketball team, the softball team — was hiding beneath my attire. But that girl was gone.

# 4

## *Conti la Monty*

AT HER JOB CLERKING AT NORWOOD HOSPITAL, MY MOTHER BE-
came interested in the work of nurses, so she enrolled in an eighteen-
month program at the Henry O. Peabody technical school to become a
licensed practical nurse. The hospital paid her tuition, and the town of
Walpole gave her a grant for child care. She attended classes all day like
we did and then shut herself in her bedroom to finish her homework,
emerging later to cook dinner. We were not supposed to bother her
when she was studying, but we found excuses to interrupt, and so she
installed an eye-hook lock, which allowed the door to open an inch. I'd
stand outside her door, my eye cocked on that sliver, spying on her as
she sat at her desk ignoring me. I'd press my mouth and nose into the
crack. *Mom, I need you. Mom. Mom.*

One day I popped the lock with a knife and barged into the room. I
found my mother crying, papers spread before her, book propped open
on her desk. "What's the matter?" I asked, forgetting whatever was so
urgent that I'd broken into her bedroom. "I can't do these fractions,"
she said. I didn't know how to comfort my mother. "Want me to help?"
She looked annoyed, or defeated. It must have been difficult to accept
that her teenage daughter was more proficient in basic math than she
was. But maybe my mother wasn't crying about fractions; maybe she
was crying about the circumstances that reintroduced fractions, those
broken-up numbers, into her life. "I have to do it myself," she said, so

I left her alone, and never again broke into her room while she was studying.

My father, whose bachelor's degree was in math, helped my mother with her homework, translating American measures into metric, into the apothecary system. In spite of tutoring from my father, halfway through the eighteen-month program my mother considered quitting. She was in classes all day, with hours of homework. She had seven children to care for, the youngest, Mikey, just four, then five. She wasn't working, so we never had enough money. We ate "plate pancakes" for dinner, the batter so thin the pancakes spread like crepes. My sisters and I urged my mother to stay in school. "What kind of role model will you be for us?" we said. She forged on, and in February 1975 she was listed in the *Walpole Times* as a graduate of the Henry O. Peabody School for Girls Practical Nursing Program, a girl at thirty-seven.


At first my mother worked for a nursing agency, driving to hospitals and nursing homes within thirty miles, and into Boston as temporary staff, usually the afternoon shift, 3 p.m. to 11 p.m. On Friday nights my father picked up Mikey. At my father's apartment, he and Mikey played with clay. My father had taken up sculpture, and his apartment was populated with half-formed figures, a bas-relief of a weary man in repose, a pony mid-trot, for which he won honorable mention in some local art exhibition. Mikey formed an army of tiny weird creatures while my father honed his bust of a woman in brown wax, sculpting and shaping as if he were creating the perfect woman, though over the years the wax woman bent and distorted.

My father came to get Mikey one Friday night when my mother was at work and her new boyfriend, Ed, was at our house. My mother had met Ed on a trip to her hometown in New York, and after that they took turns visiting each other, my mother driving four hours one way to New York one weekend and Ed making the trip to Walpole the next. Ed was the opposite of my father, like a different species of man. He was short, Italian, blond, blue-collar; my father was tall, Irish, black-haired, white-collar. Ed worked as a millwright at a nuclear power plant and was handy; he could fix or build anything. My father had a master's degree but was inept at home repair; he generally used tinfoil as his medium of choice for fix-it projects, so much so that at Christmas we gave him industrial rolls of tinfoil. Ed was a hunter, an outdoorsman. My city-bred father was useless in nature. When Mikey's pet rabbit, Marshmallow, froze overnight in its cage, my father set the bunny in front of an electric heater as if it would thaw out and start hopping.

Ed had thin lips and a long nose, close-set blue eyes. He smoked cigars — always a cigar clamped in the corner of his mouth. He wore a ring with a chunky stone and liked a crease ironed in his jeans. He was a Vietnam veteran. "He went in a private and came out a sergeant," my mother said proudly. In a hushed voice, she told me that Ed had witnessed the worst carnage of the war. "He can't eat rice," she said, a synecdoche for his whole unspeakable experience. Like Arnold Logan, the mailman, also a vet, Ed could not hold my gaze. His eyes skittered away, or he looked at the floor, as if by staring too deeply into his eyes I'd see what he was hiding — some unfathomable, lacerating sorrow.

Usually Ed left the house for an hour or so when my father was scheduled to come over, but that night he dawdled. My stomach twisted with worry as he lingered ten minutes before my father was to arrive, five minutes, and when my father walked through the door, Ed was in the kitchen holding his coat and keys. Ed extended his hand to shake, but my father frowned, looking Ed up and down, sizing him up. Ed walked past my father and out the door.

"Why'd you do that?" I said to my father.

"Do what?" He stood with his hands on his hips, still in his suit coat from work, his shirt unbuttoned, tie hanging loosely. I'd never seen my father treat anyone rudely. He rarely yelled. He was conflict-avoidant, would rather walk away than argue, but his weapon was subtle cutting remarks, a supercilious air that quietly struck you down.

"Why did you give him a dirty look?"

My father was taken aback. "He's living in this house that I'm paying for?"

"You didn't have to act like an asshole," I said.

I ran upstairs to my room. My father didn't come after me. Upstairs was almost off-limits to him. I think he was afraid to come upstairs, didn't want to trespass. Or maybe he didn't know what to say to me, or was just angry, like I was.

My father didn't see that Ed brought us crates of vegetables from his produce stand, a side business, or that he simmered a marinara while my mother was at work, that he fixed whatever was broken around the house. He'd built a wet bar in the Orange Room and a wishing well in the front yard, a stone base covered by a wooden roof, a well with no water, no purpose but to fulfill a wish for my mother. He built a deck for our pool, which made it seem almost as good as an in-ground pool.

I didn't see my father's side, that he'd been displaced from his home, that he'd become the visitor who dropped in, the interloper, not the man who slept there. My parents had agreed that they wouldn't disparage each other in front of the kids and they never did, which was why the incident with Ed surprised me. In the first couple of years after they separated, my mother and father tried to reconcile, even as my mother was seeing Ed. (Ed said he'd wait two years for my mother.) One night a few months after their separation, my parents went on a date to the movies. Around 11 p.m., after everyone else was asleep, I walked into the kitchen and caught them in a passionate kiss, my mother on tiptoe. I remember my father's words from our family meeting — "We haven't been getting along" — how that revelation had shocked me because I'd never seen them fight, but I realized that night in the kitchen that I'd never seen them hug or kiss either.

Another time my mother and father planned a picnic on a Saturday afternoon. What was it like to date a man with whom you had seven children, to whom you'd been married for fifteen years? My mother wore a sleeveless blouse and purple shorts, the hem high up on her thighs to show her shapely legs. When my father picked her up, I said, "Have fun." Hours later, after my father dropped her off without getting out of his car, she walked straight into her bedroom. I pushed open the door and found her sitting on the bed, teary-eyed. "What happened?" I asked. I couldn't imagine what might have gone wrong — something horrible if it made her cry. "I can't talk to him," she said. I didn't understand, and she couldn't explain, or didn't want to. She said she wanted to be by herself, so I left her there in her sadness.

About twice a month Ed pulled into our driveway on Friday night in an ever-changing series of fancy cars — a Toyota Celica, a shiny black Thunderbird with whitewall tires, a new car every year, which he let us drive, even the fully loaded baby-blue Lincoln Continental Mark IV with a leather interior, a huge hovercraft of a car that felt like driving a king-sized bed.

One day Sally and two of her friends and Alison and Paula and I — three girls in front, three in back — cruised the hilly back roads in the Conti la Monty, as we called Ed's car, going on "runs," getting stoned and driving Nemo Road or Moose Hill Road, roller-coaster roads with bumps and dips that momentarily floated your stomach if you drove fast enough. Stoned and suddenly hungry, we stopped at a bakery. When the woman's back was turned, Alison reached into the display case and scooped a handful of something just baked in a pan, her fingers leaving gouge marks. Alison ran out of the bakery and fell into the backseat, licking gooey batter off her fingers, slight dimples under her electric-blue eyes showing when she laughed hard. Alison was that way — taking what she wanted, grabbing, audacious and headstrong, heedless of consequences.

We were left to quickly pay for our baked goods, hoping the woman wouldn't see the tracks Alison had dug into the pan of bars or cake and catch us for . . . what, shoplifting dessert? Destruction of property? As if to restore a balance from Alison's taking, as we drove away, sick from sweets, Paula puked on the leather seats and carpeted floor of the Conti la Monty that Ed kept immaculately clean.

Ed became my mother's partner in crime. They'd walk into stores wearing old shoes or old coats and walk out wearing new ones. Ed was the kind of guy who mysteriously procured things, once a German shepherd named Max, who supposedly had gone to obedience school but clearly had flunked. The dog drove my mother crazy, and Ed took him back to New York. Later Ed brought us a pygmy goat and built a pen at the edge of our yard. The goat, Bucky, followed Mikey around the neighborhood, but its predawn bleating inspired complaints from Mrs. Wagner, so Ed gave it to some farm. A silver food van appeared in our driveway one Saturday morning, HOT DOGS in faded red paint above the service window. Ed thought we might want to fix it up and take it to beaches and concerts, but everyone was too busy or unambitious, and one day the hot dog truck was gone.

One weekend Ed towed a camper from New York that was "hot," my mother said, some sort of "crooked deal." He backed it into a corner of our driveway, which made my mother anxious. She wanted to "get it legal or get it off our property." It's ironic that my mother didn't lie to me about the camper, even though she lied to the Registry of Motor Vehicles and participated in grand larceny (since the trailer was valued at over $250), which carried a penalty of up to five years in state prison, not to mention her being an accessory to interstate transportation of stolen property, receiving stolen goods, and insurance fraud. I watched my mother back out of the driveway on her way to the Registry of Motor Vehicles, the forged documents to transfer ownership tucked in her pocketbook on the passenger seat, her head just above the steering wheel, staring at the road ahead, intent, not looking back as she drove away.

She returned a couple hours later, happy that she hadn't been caught and that the camper had legitimate plates. The crime seemed easy, just another errand to run—hardware store, bank, post office, pick up license plates at the Registry of Motor Vehicles (with doctored title), bread and milk from Fernandes on the way home. My mother painted the camper black and white and sewed gingham curtains for the windows, and she and Ed took Mikey and Ed's son, Little Ed, camping and fishing on Cape Cod. This was stealing not to support a drug habit or to get rich but to return to some middle-class norm. My mother stole an American working-class vacation.

•   •   •

Success inspires repetition, like gambling. The engine had seized in Sue's car, a Ford LTD; the car was dead but otherwise in good shape, no rust, decent interior, a high Blue Book value. I don't remember who concocted the plan, but we decided to have the car stolen and burned to collect the insurance money. Arson. Insurance fraud. I knew someone we could hire; at fifteen I had connections — Paula's brother, Duane, who was twenty. Duane was tall and strapping, like the "after" figure in the Charles Atlas bodybuilding ads, with a thicket of auburn hair and light-brown eyes like Paula's. Duane said he'd do the job for $75, and we set a date. Since the car didn't run, my mother and Ed towed it with chains to the parking lot of the Walpole Mall.

That night, Sue and her boyfriend, Jeff, ate at a pub in the mall. After a certain period of time passed, they walked out to the parking lot to discover that Sue's car had been stolen. She reported the crime to the police. The next day the cops told us they'd found the car at the gravel pit on Route 1, a mile from the state police barracks, where Duane had towed it while Sue and Jeff were in the pub and then had set it on fire, though only the trunk and rear end burned, which caused some delay in processing our claim. Eventually we collected the insurance payout of a couple thousand dollars.


My mother found a full-time job at Pondville Hospital, run by the state of Massachusetts, a two-story brick building hidden in the woods at the end of a winding driveway, almost directly across the road from Walpole Prison. Pondville mainly treated cancer sufferers and the occasional prisoner from Walpole with minor injuries, superficial stab wounds. My mother brought home stories of her "end-stage" patients, their sorry states, how she held their hands and comforted them while their lives slowly ebbed. For me, her only kid who smoked cigarettes, she described in graphic detail the halting, gasping breaths of lung cancer patients, their phlegmy spasmodic coughs. She threatened to drag me into Pondville to witness their suffering, their slow agonizing dying, which surely I'd face if I didn't quit smoking, using the hospital as an object lesson, as she had the prison. I scoffed. I hadn't been deterred by the antismoking film in home economics in junior high, by the image of an emaciated woman puffing a cigarette through a hole in her throat and speaking in a buzzy, robotic machine-voice.

Pondville Hospital was one of five state institutions within an eight-mile radius of my house, places that hid away the less fortunate, the misfits. Along with Pondville, Walpole Prison, and Norfolk Prison, the Wrentham State School for "mentally retarded and wayward children" (the "feeble-minded," they were called when the school opened in 1910) was a mile and a half away, a two-story building set on a granite foundation, with five hundred acres of agricultural land, where in the early years developmentally disabled boys learned the manual labor of farm work. In winter I practiced with the swim team at Wrentham State's indoor pool, the water heavily chlorinated because the mentally retarded kids peed in the pool, my eyes red and burning after an hour in the too-warm water.

During the years I swam there, the school was under investigation for its deplorable conditions, which a judge called a "disgrace and a nightmare." A few years later the school's certification was revoked because it failed to meet minimum standards of care. Decades later it was revealed that in 1962, to test the effects of nuclear fallout, professors from Harvard Medical School had given high doses of radioactive sodium iodine to sixty-three "mentally defective" children at Wrentham, ages one to eleven, without permission from their families — a crime as deplorable as any committed by a Walpole inmate, though nobody was ever held accountable.

The fifth state institution, about four miles north of the prisons, was Medfield State Hospital, opened in 1892 as the Medfield Insane Asylum, which housed up to 1,500 mentally ill and drug-addicted residents. Spread across nine hundred acres were fifty-eight buildings, many with elegant architectural details: dentil molding, Doric columns, arched windows, slate roofs. The administration building, where I picked up my mother from float shifts she worked to earn extra money, looked like a country mansion.

Inmates from Walpole and Norfolk Prisons worked "trusty" jobs at Pondville Hospital, Medfield State Hospital, and Wrentham State

School, probably because they were cheap labor, earning less than a dollar a day. Perhaps, too, someone in power thought that working with psychiatric patients or disabled children humanized prisoners. The "retardates," as the newspaper called the Wrentham kids, wouldn't fear the inmates, wouldn't understand that their caregivers were murderers, rapists, thieves, drug dealers. A Norfolk inmate, Leo Hurney, a trusty who worked forty hours a week attending psychiatric patients at Medfield State, said, "I'm still a humanitarian even though I'm a con."

At Pondville Hospital prisoners worked as janitors or in the laundry. One night an inmate in his midthirties asked my mother on a date; he was going to be paroled soon. She asked the man why he'd been sent to prison, and he told her that he'd killed the man with whom his wife had cheated. At least he was honest. My mother thought he seemed nice, but she declined his offer, nervous about the idea of dating a murderer, and anyway she was seeing Ed.

The trusties wore special uniforms so that employees and patients could differentiate the inmates from the regular janitorial staff, like my brothers, Patrick and Michael, who worked at Pondville part-time in high school. Michael's supervisor, Jerome, was a "good guy," even though he'd killed a man over a drug deal; he and his brother had both been sent to Walpole Prison. Patrick worked with an inmate named Tony, a huge, toothless African American man in his midsixties, who at nineteen supposedly killed three men in a bar fight, like something out of a Grimms' fairy tale. Tony had been in prison almost half a century. He believed he'd never be released, but at least he was out for a few hours every day, mopping floors and emptying trash. He and Patrick smoked joints on break, and Patrick gave him pot to smuggle inside. Tony snipped the fingers off surgical gloves, stuffed them with pot, knotted the packets, and hid them in his rectum.

Some of the trusties gave off a menacing vibe, like Reg, in his fifties, convicted of murder, who Patrick thought might stab you if you turned your back to him. The janitorial staff—Patrick and a couple of other teenage boys with after-school jobs, and the inmates—played cards in the break room, the boys awed and a bit afraid of their criminal co-workers.

One day when I picked up my mother after her shift at Pondville, as we walked past a man in the hallway, she whispered, "He murdered someone." I looked back to watch the murderer swish his mop across the linoleum. His face showed nothing of a killer. I didn't know what a murderer should look like, but he shouldn't have a sunken back and white hair and a big belly overhanging his belt. I thought a murderer was a specific type you could identify. I wouldn't have thought that a murderer would be Mikey's classmate's father, Leroy Chasson, whose wife — dressed as a nurse and wielding a .45 — broke him out of Norwood Hospital after he'd purposely injured himself in Walpole Prison. The Chassons fled to Colorado, where they lived for seven years until *America's Most Wanted* exposed them.

Or a druggie I saw around town in Walpole, Woody, who in a few years would kill his best friend, Billy "Lighty" Lightner, and wind up in Walpole Prison. Or one of Patrick's friends in high school, who slashed a kid's throat with a broken bottle during a fight at a party one night. None of the bystanders helped the dying kid, because they didn't want to get in trouble, so he bled out before the cops arrived.

When I needed to borrow my mother's car, I drove her to work and picked her up after her shift. Often I saw clusters of people in the prison parking lot carrying signs or milling around, and I wondered what was happening — perhaps prisoners' rights groups protesting inhumane conditions on Block 10, the segregation unit, or members of the media waiting to photograph the return of some escapee, or a famous prisoner shuffling in ankle chains. Maybe the commotion was correctional officers on strike.

Throughout the 1970s there were frequent clashes between guards and the prisoners' rights group, or between guards and the administration, especially John Boone, Massachusetts's first African American corrections commissioner. In the early 1970s, Boone tried to introduce progressive measures, but in protest of his liberal policies, guards walked off the job. During the guard strike from mid-March to mid-May of 1973, except for state troopers posted at the perimeter, the inmates ran Walpole Prison, a rare time of interracial cooperation and calm. A pool of 550 citizen volunteers, including the Walpole Jaycees, served as observers, stationed on the cellblocks twenty-four hours a day.

The decade of the 1970s was the most violent and troubled in Walpole Prison's history, emblematic of prisons nationally. In 1973 in the United States there were ninety-three riots for every million prisoners; in 2003 there were fewer than three. In 1973 there were sixty-three homicides per one hundred thousand prisoners; in 2000 there were fewer than five, even though the incarceration rate had quintupled since 1970. In Walpole Prison between 1971 and 1973 there were two major riots, a guard strike, fires, work stoppages, shakedowns, lockups, widespread drug use, and four superintendents in three years (one said to have suffered a nervous breakdown).

There were dozens of stabbings and violent inmate deaths. In 1973 a *Boston Globe* headline read, "Walpole Tops Nation in Inmate Killings." The inmate murder rate at Walpole Prison was more than three times higher than in San Quentin, more than four times higher than in Marion Federal Penitentiary in Illinois. The final murder of that murderous year was in November, when Albert DeSalvo, the Boston Strangler, was stabbed sixteen times in his chest while he slept in his supposed protective-custody cell.

The violence continued throughout the 1970s, with vicious killings. One inmate was stabbed fifty-seven times in his cell. Another was beaten to death by his cellmate (who'd bludgeoned to death his former cellmate with a toilet seat). Another prisoner, Robert Perotta, was killed in a gruesome manner, strangled, then castrated. In 1974 twenty-two-year-old John Kelly, the brother of a friend of Sue's, had been a guard just four months when he and another guard were taken hostage, bound and gagged for twenty-four hours, then released unharmed. After the media had been barred from the prison, the inmates took hostages, because they needed a way "to speak to the people," their statement said.

Overcrowding was blamed for the trouble; the prison was designed to hold 566 men, but by 1976 there with 693 inmates. State legislators toured Walpole and found rats in the cells, debris all over the floors, and conditions so hazardous that they worried about a hepatitis outbreak. An officer recruit who visited Block 10, the segregation unit, said, "There was shit, garbage, everything thrown all over the place. Human shit. And it stunk." Another officer who worked Block 10 saw "garbage [knee]-high on the floors — feces, urine . . . an inch of slime on the floor so when it was wet outside, you'd slide down the corridor." Governor Michael Dukakis was urged to shut down Walpole Prison. I was oblivious to much of the turmoil, except when the trouble boiled over the walls into the front lot, where we used to park for the Hobby Shop.

Sometimes when I passed Walpole Prison on my way to Pondville, I'd see a lone figure sitting inside the bus stop, a small brick shelter across the street from the prison, perhaps a just-paroled ex-con, someone without friends or family to pick him up so he sat in the tiny hut staring at the cement walls across the road, outside but still too close as he waited for the Greyhound that traveled Route 1A, picking up riders at the prison stop and then in Walpole center, where my friends and I used to board to shoplift at the Dedham Mall or sometimes rode to the end of the line, Forest Hills, where we caught the train to Boston.

What did it feel like to be released after three or five or ten years of imprisonment, fifty bucks of gate money in your pocket, a set of civilian clothes, only to sit across from the prison at the bus stop waiting and waiting for a ride out of Walpole?

# 5

## *Clarity and Logic*

IN THE FALL OF MY SOPHOMORE YEAR, I TOOK AN ENGLISH CLASS called "Clarity and Logic," an accelerated course for students who'd been recommended to skip the required tenth-grade English. None of my friends were in the class. In "Clarity and Logic" we read "The Loneliness of the Long-Distance Runner," by Alan Sillitoe. The narrator, Smith, was a seventeen-year-old "outlaw bloke" in a borstal, a youth detention center in England, where he was recruited by the warden to run a long-distance race to win a medal for the prison. "I didn't mind it much," Smith says, "because running had always been made much of in our family, especially running away from the police." I wonder whether our teacher, Mr. Hubbard, a handsome man who played rugby (which might have explained his chipped front tooth), chose this story because we lived in a prison town, or maybe just because the protagonist was a teenager, like us.

In class Mr. Hubbard asked why Smith, a gifted runner, purposely lost the race. He waited, staring at our blank faces, hoping someone would respond, these smart kids who would graduate at the top of their class, become salutatorian, valedictorian. Maybe these kids couldn't identify with the protagonist, a delinquent, couldn't imagine intentionally losing a race they could easily have won. I'm sure I was the only one in the class who'd ever done anything wrong, who'd earned detentions, who'd broken laws. The spirit of that delinquent boy lived in my bones. I could see that his act was about freedom — that he chose to lose, to forfeit the trophy and the approval of his keepers, for a larger prize, autonomy, to be true to his spirit. He rebelled, defied authority, exercised power in a powerless situation. His reason was clear and logical. I raised my hand.

After that, Mr. Hubbard would look to me for answers in class discussions, but one night I smoked angel dust and again the next day and the next until it became a habit.

> November 10, 1975 — No school tomorrow. Went on runs all night with Paula, her brother Duane, and his girlfriend. We bought a gram of dust. That's the first time I ever got dusted. Dust is heavy but it screws up your brain cells.

After that night, what I recorded in my diary nearly every day was what drugs I took and with whom, the beginning of a long, slow project of erasing myself, or trying to. There were moments when my two worlds mingled, when I wrote that I'd landed my backflip in gymnastics, then got wasted on angel dust that night, or that I bought angel dust with money I earned babysitting. In Mr. Hubbard's English class, I'd sit in the back row, stoned or dusted, and stare out the window at the kids in the parking lot. I stopped doing my homework, and when Mr. Hubbard looked to me for an answer, I was silent. I'd lost all clarity, all logic.

In Walpole in the mid-1970s there was a dry spell for marijuana, which was likely a consequence of the war on drugs launched by Nixon, and continued by President Ford; during this campaign, the U.S.-Mexico border was tightened and U.S. helicopters sprayed the herbicide paraquat on Mexico's marijuana fields. In that window of time — after the crackdown on Mexican marijuana but before enterprising Americans cultivated high-quality pot and it was easily available again — angel dust, a made-in-America drug, flowed into Walpole. That year, 1975, was on the steep incline of high school kids using illicit drugs, which peaked in 1979 and then fell dramatically, remaining lower since then.

Angel dust is phencyclidine (PCP), synthesized by Parke-Davis in 1956 as a surgical anesthetic. PCP was first tested on monkeys, who were so serene on the drug that Parke-Davis named it Sernyl. In later studies, though, monkeys quickly developed a tolerance for PCP, self-administering higher and higher doses, sacrificing food in favor of the drug. Over a period of several weeks the monkeys took such high doses that they were "frequently found lying on the floor of the cage in awkward positions, briefly raising themselves to press the lever only to fall back down to the floor." In one experiment, by the end of the trial period the monkeys were using four times the original amounts of PCP.

In spite of these alarming results, drug trials progressed from monkeys to humans, at first a few dozen patients at Detroit Receiving Hospital, many of whom woke from surgeries unable to feel their limbs or wondering where their heads were or why the surgeon had turned into "a vampire bat with a ten-foot wingspan and four-inch claws." Patients with strong religious beliefs thought that they were being "carried up into the clouds" by God.

For two more years Parke-Davis experimented on humans, mostly hundreds of prisoners in Jackson, Michigan, the largest prison in the world then, with some five thousand inmates; Parke-Davis built a lab inside the prison. People continued to experience bizarre, near-psychotic reactions to PCP, so finally, in 1965, Parke-Davis abandoned its plan to market the drug for people. The FDA approved PCP for veterinary use, mainly to tranquilize large animals. Phencyclidine, renamed Sernylan, was used to stun horses and cows weighing a thousand pounds to a ton. That was the drug that I was putting into my one-hundred-pound body.

In Walpole I'd seen angel dust only as dried parsley onto which PCP had been sprayed or baked. We smoked dust in a pipe or rolled it in a skinny joint with Zig-Zag papers, tissue-thin like the shed skin of a snake. No one I knew referred to PCP as anything other than angel dust, or dust, or killer D for an extra-strong batch. Angel dust highs felt pleasant and dreamy at first. Driving around that first night I smoked dust, Duane's car seemed to glide into the parking lot at Friendly's, where we stopped to use the bathroom. Inside, kids from school were lined up on stools at the counter, eating ice cream: my former friends. When I walked over to say hello, the linoleum floor turned spongy, as if I were in that commercial for Hush Puppies shoes, a man walking along a springy sidewalk. When I was a kid, that commercial had made me desperate for a pair of Hush Puppies so that I, too, could bounce on foam rubber sidewalks.

Angel dust made you feel spacey and untethered, like Major Tom, *floating in a most peculiar way.* We called dustheads "space cadets" or "space shots." On angel dust you lose proprioceptivity, your sense of your body in space. Angel dust causes ataxia, a neurological disorder in which people lose balance, depth perception, and coordination of bodily movements. In Walpole center you'd see kids doing the dust-walk down Main Street: a stilted, jerky high-step, like a marionette, as if they were lifting their foot over a curb with each step, like in Fly, a game my siblings and I played as kids. We'd take the mirrors from the walls in our bedrooms and hold them perpendicular to our chests like trays — just below our chins, so that the floor was blocked from view and the ceiling reflected in the mirror — and we walked around the house like flies walking on the ceiling, high-stepping over doorframes, careful not to trample light fixtures, moving across a stark white land-scape like the astronauts on the moon.

Angel dust causes confusion, agitation, euphoria, and difficulties in abstract reasoning, attention, and concentration. Dust causes "autistic dreamy states" and "cloudy delirium" and, at high enough doses, cata-tonia. Psychologically, angel dust causes feelings of apathy or estrange-ment, feelings of nothingness, or a preoccupation with death (*medita-tio mortis*). A friend of Duane's swore he was going to quit dust after he thought he was dead for three days.

Dust distorts the size and shape of your body. Paula and Alison and I smoked angel dust one Saturday before going to the high school foot-ball game. We walked along the chain-link fence near where it abutted woods, on the far side of the field, where someone had cut a hole to sneak in without paying. The angel dust transformed us into midgets, like Alice with her mushroom or Alice with her pills — one that makes you small. We were suddenly tiny, like in that Arabian Knights cartoon I loved as a kid, in which a turban-wearing man clapped his hands and said, "Size of a mouse." Dusted, we shapeshifted to dwarves, but we couldn't figure out how to climb through the hole in the fence. We couldn't think. The fence puzzled us, and we walked back and forth along a section of the chain link, like a neurotic polar bear I saw once at Franklin Park Zoo in Boston, missing one eye, lumbering senselessly.

Early in the phencyclidine tests on humans, a group of psychiatrists compared the reactions of people on PCP to those of four other groups: people who'd taken LSD (legal until 1967), people who'd taken amobarbital (a barbiturate, basically a sedative), schizophrenics, and a control group. The researchers looked at how the drugs affected two specific functions, sequential thinking and symbolic reasoning. For the sequential thinking test, the subjects had to count backward from 100 by multiples of 7. The average score for the control group was about 90 percent correct. People on LSD scored slightly lower, 78 percent correct, while those on the amobarbital scored slightly higher, 91 percent. The score for people on PCP was 32 percent accurate, close to the scores of the schizophrenics. PCP, the researchers wrote, was "psychotic-mimetic" — it mimicked responses of people suffering from psychoses.

To test symbolic reasoning, the subjects were asked to interpret the meaning of five proverbs:

> The proof of the pudding is in the eating.
> A rolling stone gathers no moss.
> A drowning man will clutch at a straw.
> One swallow doesn't make a summer.
> Great bodies move slowly.

The average correct score for the control group was 5.1 out of a possible 10. PCP subjects scored less than 1.0. Even schizophrenics beat the PCP group, scoring 1.5. The people on PCP were utterly confounded by the proverbs. Trying to explain the meaning of "the proof of the pudding is in the eating," one subject said, "Good food is worth eating," while another said, "The eating of the pudding will prove itself." One subject's response epitomizes the PCP befuddlement that I remember. When asked, "What does this proverb mean?" the subject said, "What does *mean* mean?"

Angel dust was not a drug that increased sensory experience or made the world seem funny, like pot or acid. Angel dust killed desire for food or sex, bodily hungers. Dust incapacitated thought and speech; your brain no longer functioned, and sometimes you couldn't form basic words. Cognitive processes shut down, a torpor of the brain. PCP was a *dissociative anesthetic,* a new category created just for this drug. Angel dust was the perfect drug for me; I wanted to anesthetize myself from pain, from sorrow; to dissociate myself from myself. Detached from my body on angel dust, I erased myself. I was no-body.

One of the clinical symptoms of angel dust is depersonalization, as if you are in the process of unbecoming, becoming *not* a person. If the drugs of the 1960s — marijuana, LSD, even heroin — were about tuning in, about pleasure and euphoria, angel dust in the 1970s was about zoning out, reflected in the music of the decade, like Led Zeppelin's "Dazed and Confused," the twanging guitar notes bent by the wa-wa bar — *Been dazed and confused for so long, it's not true.* The song is about love, but it meant something different to us. Or Pink Floyd's "Comfortably Numb" — *There is no pain you are receding.*

When I was very small, if I fell off my bike or knocked my funny bone and it tingled with pain, I'd cry to my father, who always said, "Rub it off, it's all right now, just rub it off." He'd place his big warm hands on my elbow and rub my skin, and then I'd continue to rub the spot with my hand, and somehow after a few minutes my arm no longer hurt. It was just that easy to rub away pain. As a teenager, my pain was inside, a twisting feeling behind my sternum some days, a knot of anxiety in my stomach. I was always a nervous kid, so worried about being late for school that I'd cry if my mother delayed me to sew a loose button on my blouse, then I'd race out of the house, my chest numb with panic, certain that I'd missed the bus, especially when I didn't see any kids at the bus stop, only to realize that I was the first one there.

I smoked angel dust to rub off psychic pain — anxiousness, self-consciousness, alienation, confusion — in that simple way that had worked when I was kid with a bruise or a bump. At the time I didn't understand my motivations for taking a drug that was literally an anesthetic. I thought I was just experimenting, exploring the world as I had in childhood, pushing boundaries further and further, like the time I was nine and I rode my bike along the tracks so far one day, curious about what was ahead, what lay just beyond, so I pedaled and pedaled, stopping abruptly when I saw the sign that announced I was entering Medfield, a whole other town, feeling a bit scared then by how far I'd gone, how easy it had been. With angel dust, I'd be too far gone before I realized I'd gone too far.


I smoked angel dust with Nicky Osborne, my new boyfriend in tenth grade. He wasn't in any of my classes, but I always picked him out of the landscape, which seemed uncanny. Slender and slight, with thick, curly, wheat-blond hair, he seemed to glow, with his pale skin and blushed cheeks. I loved him from afar, noting in my diary how beautiful he was. As if willing it, I'd find myself standing in the same circle of kids passing a pipe, the proximity to Nicky sparking an electric hum in my body, like the time I'd stuck a knife in the toaster when it was plugged in. He was quiet, not rowdy like many boys; he had a gravitas to his presence that belied his slight physique. I thought he was too beautiful for me, the delicate oval of his face, his hazel eyes. I was shy around him, as I hadn't been before with boys. I didn't dare speak, worried I'd say something uncool. There was a high school dance at Blackburn Hall one Friday night. I drank some beers and smoked some pot but "didn't get very buzzed," I wrote, though Paula was "really blown." I took Paula outside for air, holding her long hair away from her face as she vomited in the bushes and, unfortunately, all over her coat. I saw Nicky outside, wearing work boots that looked too big and powder-blue thin-wale corduroys, which everyone wore, boys and girls. He wore a puffy down jacket and a wool hat that brought his head to a point, loose curls bunched around his collar. I saw him, but he didn't see me. "Nicky was so fucked up he couldn't walk or see where he was going. He didn't even know where he was." Nicky being *so fucked up* did not deter me from liking him; getting wasted was the norm. Getting high was what all the cool kids did, all the beautiful cool kids living on the edge.

In high school the division between jocks and freaks was like a partition in a country. It began in junior high, when tough girls kicked the shins of cheerleaders walking down the halls and then the conflicts escalated into fistfights after school. This was the age of equal opportunity, girls acting like boys. Nancy Morris versus Debbie Drake. Karen Hoover against Sandy Bartko. "Brenda Stokes and Janet Ballard had a fist-fight after school today. I think that's sick," I wrote. I called the fighting sick probably because I felt queasy whenever I saw someone strike someone else. In ninth grade I straddled both camps, walking the line between jocks and freaks like walking the balance beam in gymnastics.

The jock-versus-freak fights between girls continued in high school, at least until my sister Sue, who was president of the student council, negotiated peace, or détente, between the camps, sitting down with representatives from both factions. The violence stopped, but the divide was still palpable. There were territories in the back parking lot behind the school, on the far end the freaks and nearer to the school entrance the jocks, if they dared stand outside at all. Mostly they stayed in the cafeteria before school or at lunch. The boundary line seemed impenetrable; you had to choose, and there was no crossing back.

Stepping off the bus into the high school parking lot felt like passing through a squeeze chute, self-sorting by turning right or left, taking a stand by where you stood. I turned right, choosing to stand on the freak side, the wild side, outside.

One day at school Nicky saw me at my locker and said, "How ya doin', Maureen?" I turned to see him walking backward in the hall, wearing his ubiquitous hat, carrying no books, grinning a wide grin, his perfect smile. He turned before I could reply. I was "so shocked," I wrote in my diary. He called me that night, and we planned to meet that weekend. "How unreal," I wrote, my dream coming true. That Saturday night, Paula, Loretta Petty, and I walked down Robbins Road toward the junior high, always deserted after school hours. We spied two figures and walked toward them, then stood in an awkward circle with Nicky and his friend Vince Gentile, a handsome Italian boy with tea-saucer brown eyes. I saw right away how dusted they were; neither spoke. Every few minutes Nicky or Vince said, "Whoa," as if they were on some wild ride, as if they could feel the planet spinning beneath their feet. Their minds were blown, like a circuit, a bulb burned out. That's what we called druggies who hung around downtown: burnouts. Wayne Kosinski, Rod Tyler, Billy Lightner.

After a while Vince smiled lopsidedly. "Say a few syllables," he slurred, a line from *The Three Stooges,* the episode in which the Stooges were prisoners in jailbird garb, numbers emblazoned across their chests like on my mother's Halloween costumes. The warden was unconscious. "Speak to me, warden," Moe said. "Say a few syllables. Utter a few adjectives." Loretta whispered loudly to me or to Paula, talking about Nicky and Vince right in front of them, giggling, crunching Doritos, leaning in to gossip with her sharp Dorito breath. I found her irksome, the impolite, immature whispering, the annoying crackle and crunch, her suffocating cheese breath in my face. I recorded the sentiment: "Loretta was really bugging the shit out of me and Paula last night, trying to act cool. She's a fucking asshole." I'm sure I meant that Loretta was trying to act cool, but I can read the subtext — that it was Paula and I who were trying to act cool, Loretta foiling our efforts. I worried about appearing cool to a boy who'd lost his capacity for speech. I considered Loretta impolite, not the boy who showed up wasted on our first date.

When I stepped off the bus at school the next day, Loretta and Chrissy Mayer and Lisa Summers and Laura Pagano surrounded me like a posse, whispering, too eager. *What happened with Nicky? Tell us, tell us.* Nicky was standing twenty feet away with Vince, smoking a cigarette. I was embarrassed by my friends acting girlish, these friends I'd soon abandon, with their schoolbooks clutched to their chests, their feathered haircuts, their fear. I didn't want to tell them anything. What could I say? That Nicky was so dusted he couldn't speak? I brushed past them. "Leave me alone," I said.

The next Saturday night Nicky and I sat on the bleachers behind the town pool, where kids were partying, where the year before I'd practiced with the swim team, riding my bike at 7:30 every morning for an hour of laps, then an hour of junior lifesaving. In the lifesaving course I'd aced the written tests, but during the final practical exam the 175-pound lifeguard wriggled free from my cross-chest carry and I failed the class. I cried all the way home on my bike. I could save someone only in theory.

Nicky's sister, Andrea, a year older than Nicky and pretty, as he was, walked over to us and asked to borrow a few bucks to pay someone back, she said. Nicky and I scrounged up the money, but after she left he worried. "She'll probably buy dust," he said, concerned for his sister, though not for himself. Nicky apologized for being so dusted that first night we met, said he'd been "psyched" to see me but that he'd ruined it. "That's okay," I said. At times I'd been *so dusted,* too. Nicky's friend Tracey walked by and said she had an ounce of dust, twenty-eight grams, enough to keep several people dusted for a couple weeks. "I'm quitting," Nicky told her. Tracey looked skeptical when Nicky turned down her offer, and based on how often I'd seen Nicky dusted, I was, too. When Tracey left I said, "When you go home tonight, I bet you'll smoke dust."

"All I'll do is think about you like I did last night," Nicky said, and my heart tumbled a bit. He asked me if I wanted to go out with him "for as long as it lasts" — a curious proposition, agnostic but honest. Shivering in the sharp November air on the cold aluminum bleachers, I said yes, and then there was nothing but Nicky and me kissing.

Most often Nicky thumbed a ride or walked the few miles to my house or we met in the town center, which was halfway. One night Nicky and I and Vince smoked dust and wandered around downtown for a while and then hitchhiked home. The boys both lived in the opposite direction from me, so they waited on one corner while I stood on the perpendicular corner. A police car stopped and I watched Nicky and Vince climb in the cruiser and drive away. I was alone now at the stoplight, my thumb out. A few seconds later the cop car pulled up in front of me, and Nicky opened the back door, grinning. "Get in." He'd cadged a ride for me, too, from Good Goin' Gus, a cop sympathetic to teenagers. When dozens of kids gathered at Friendly's on summer nights, Gus would screech into the parking lot in his cruiser, roll down his window, and yell, "Get out of here, the cops are coming," as if he were not one of them. By the time the next cruisers arrived, everyone was gone or leaving.

After Gus dropped off Nicky and Vince, he raced across town as if worried he'd be caught giving me a ride, as if it were not his job to prevent a crime that might happen to a drugged-out girl hitchhiking alone. A half mile from my house I said, "This is good." I appreciated the ride, but I wanted out of that cop car.

Good Goin' Gus peeled away and I started to run. I took huge leaps across driveways as if I were wearing PF Flyers, which as a kid I had to have because the commercial promised you'd "run your fastest and jump your highest" in PF Flyers' vulcanized rubber soles. Even Jonny Quest rescued Race Bannon from burning lava, running like the wind in his PF Flyers. That night I was convinced I was leaping ten feet with each step, like when I won the long jump at the Lions Club track meet in fourth grade, 10′7″ penciled on the back of the blue ribbon. I ran and leaped, bounding like a superhero, airborne: Velocigirl.

Someone looking out his window that winter night would have seen a skinny girl running along the sidewalk in an awkward gait, passing quickly through pools of light from streetlamps, through shadows, past the cemetery, past the projects, then Barnes's store and Delluci's field, where I used to hunt for snails on the underside of milkweed pods, past the dead-end sign at the top of my street, past the darkened houses of my neighbors sleeping their sound or troubled sleep.

In my room I fell into bed, and as happened often after I smoked dust, my mattress spun and lifted off its frame and flew right out of my bedroom window and I whizzed around in black space on my flying-carpet mattress like in that kid's book *The Magic Bed*. But I couldn't land, couldn't come down, and so I floated through the night until somehow I fell asleep.

The next day was a Thursday, and my diary began, "Got dusted with those guys again this morning." Dust in the morning, dust at night, dust at school. We smoked dust in the girls' rooms, in the school parking lot, in class. It was easy to smoke angel dust in the back of Mr. Hood's chemistry class, the irony lost on me — angel dust the product of an amateur chemist in some home lab. Hiding in the back row at the waist-high lab tables, I'd put a few flakes of angel dust in a small pipe. When Mr. Hood turned to write on the chalkboard, I'd light the pipe with a Bic lighter. A match would have released sulfur, but the lighter was odorless. I'd inhale deeply, sucking the pipe until there was nothing left on the screen but white ash. Angel dust was not oily like marijuana, so there was no cloud of smoke.

One day Vince Gentile walked into our social studies class, "America in the Twentieth Century," fifteen minutes late, high on angel dust. He banged open the door to the classroom and the conversation stopped as everyone watched him blunder in, bumping into desks, staggering to his seat. Miss Geoghegan stared through her huge owlish glasses as

Vince dragged his desk 180 degrees around to face me, his back to her. "What's up?" he slurred, as if we were outside in the parking lot. I tried to shush Vince, signal him to look up front. Miss Geoghegan called his name several times, but he didn't see or hear her. He had no idea where he was, what the context was: a class in high school. Earlier that term Miss Geoghegan had taught us that the average number of kids in the American family was 2.2. "How can you have a point-two kid?" I'd said. The class laughed and Miss Geoghegan smiled. "Oh, you!" When Vince Gentile stumbled into class fucked up on angel dust that day, he was a point-two kid — a kid not all there.

After failing to get Vince's attention, Miss Geoghegan walked over to his desk and ordered him to leave. Somehow Vince understood. He pushed his desk back, swiped the hair off his sweaty forehead, and walked into the hall and then I don't know where. This was Walpole High School in the mid-1970s; this was America in the twentieth century.

As if to presage my own year ahead, President Ford in his 1975 State of the Union address told the country, "We face troubled times." By the time 1975 was spent, instead of recording in my diary a list of presents I received for Christmas, as I had in prior years ("I got 3 shirts and 2 socks and 1 pants"), my diary entries became litanies of drugs I consumed. Christmas used to be a day of opening gifts, trying on new clothes, eating too much chocolate, and visiting my grandparents, who presented us with our annual $10 bill in a money-holder envelope. Instead, Christmas Day became this:

> December 25, 1975 — Me, Paula, Alison, Adrian and Carl went over Nicky's house. Got dusted and fried.

My friend Terry Littlefield and I went to gymnastics practice dusted one day, before I quit the team later in my sophomore year. I hadn't smoked enough dust to be incapacitated, just to feel hollow-boned, light and bouncy as I leapt around the mat pretending to flip and twirl, turning in my airy weightless body made only of dust, floating like an angel. I was not actually performing gymnastics, just spinning and jumping like a deranged ballerina. It was dark outside, so I couldn't see Nicky watching me through the window. He called me later, said I was "nutty," which I took as a compliment.

Dust took me out of 1975 and followed me into the next year, 1976. My last diary entry in 1975 was a sad one:

> Dec. 29, 1975 — I got my back handspring by myself. Met Nicky downtown. Got dusted and blown away.

Where were my siblings that winter when I started smoking angel dust heavily? Joanne, a year younger than I, tried dust once but never again for the fright she got from seeing people transformed, their faces slack, eyes vacant. One night when I was dusted, Joanne and her friend Maggie, who were straight, urged me to drive faster. I felt like I was speeding, but when I looked at the speedometer, I saw I was driving 10 mph in a 45 mph zone. Neither Joanne nor Maggie had her license, so they couldn't take the wheel.

My sister Sally, a year older than me, was a gifted artist, winning prestigious state awards in high school. Perhaps because she spent hours lost in the dream state of drawing and painting, she liked being dusted at first, the "singe" in her mind, she called it, when her head "leaped over from being normal into that weird other-person-ness and everything glowed and hummed." Once when Sally was dusted, the other-person-ness was an other-*thing*-ness, when she thought that her head had turned into a head of lettuce. Another time when Sally and her friend were dusted, it took them two hours to cross a street.

Sue smoked angel dust, too, with her boyfriend, Jeff, who distributed grams for the big-time dealers in his neighborhood — Billy Lightner, Wayne Kosinski — just enough to support his habit. That Sue smoked angel dust might have surprised some of her classmates, and certainly her teachers. In her senior year she was the president of the student council, a homecoming queen candidate, an honor roll student.

December 28, 1975 — Sue picked up Jeff and took me and Nicky on dust
runs. Me and Nicky went in on a $10 gram. Got stuck in a snow bank.
Drove around getting blown away until midnight, then Sue dropped
off me and Nicky.

I remember that night driving around with Sue and Jeff, scrunched
with Nicky in the backseat, feeling so cool hanging out with them. Jeff
was like a part of our family, an older brother, giving us rides, babysitting Mikey if Sue was working. He was lantern-jawed, with a wide
smile and soft blue eyes, a long dishwater-blond ponytail. He was tall
and broad-shouldered, with a regal bearing, though I sensed a wound-
edness about him. He was a talented musician with plans to enroll in
Berklee College of Music in Boston, saving money by working full-
time in the kitchen at Walpole Prison. One day a prisoner asked Jeff to
smuggle angel dust inside, but Jeff declined. "You're in here, and I'm
out there, so why would I do that?" The inmate looked at Jeff menac-
ingly after that, and, fearing for his safety, Jeff quit.

That night Jeff drove Sue's car, a green Ford Maverick, which she'd
purchased with the insurance payout from the car we'd torched. On
Route 1A, the main drag, the car spun out of control on the slippery
road, everything in slow motion as we pirouetted, lodging in a snow-
bank facing oncoming traffic. Red lights flashed outside the window,
which I thought were for us, but the police cruiser whizzed past. Jeff
shifted the car into reverse and somehow maneuvered it back onto the
road in the right direction, and we drove away.

Were my youngest siblings aware that I was smoking angel dust? Mikey
was just five that year; he would still have believed in Santa Claus. My
father slept over on Christmas Eve for many years, stretched out in the
living room couch so he could be with us in the morning to open presents. That year, instead of going to midnight Mass on Christmas Eve,
as I'd told my mother, Sally and I smoked angel dust, sitting in Timmy
Carroll's truck outside Blessed Sacrament Church, where I'd made my
first confession, my first communion, my confirmation at fourteen.

When Sally and I arrived home at 2 a.m., my mother was "rippin'," I wrote in my diary.

Barbie was eleven, and if she knew about dust, it was from observing her older sisters, learning what *not* to do. Barbie was nine the night Sally came home drunk and collapsed on the kitchen floor. Barbie overheard my mother say, "I hope she's not pregnant," and for a while Barbie thought that you got pregnant from drinking alcohol. Barbie had a seemingly innate sense of self-discipline. In the noise and confusion of our house, every night Barbie sat at the dining room table diligently finishing her homework, and then promptly at nine she went to bed, the same steady determination she'd applied to painting the pool posts with that nasty epoxy when she was eight.

Patrick, at twelve, was running wild with his friends, smoking pot and drinking before he hit seventh grade, later thrown out of eighth grade, probably for destroying seats on a bus with a kid from the Farm. I didn't think Patrick smoked dust, though I assumed he knew I did. One weekend my father took Patrick and me skiing in New Hampshire — my father had taken up the sport at thirty-eight — and I asked if I could bring Nicky. I borrowed Sue's car and gave the keys to Nicky, who didn't have his license yet, but he wanted to drive and I didn't. At some point we hit a snow squall, large fat flakes colliding with our windshield, Nicky and I encapsulated, snow swirling like we were inside a snow globe. We managed to find the motel with no cell phone, no GPS, just a map and some directions from my father. We all shared a room, Patrick and Nicky in one double bed, me in another, my father on a fold-out couch.

That night in the basement rec room of the motel, Nicky and I smoked angel dust, finishing just as Patrick came down, the acrid smell lingering. We shot pool, the balls cracking sharply in the quiet basement. When we walked back to our room, Nicky whispered, "Sleep on the edge of the bed." I looked at him, puzzled. "So we can hold hands," he said, but sleep relaxed our arms and we floated into separate dream worlds. The next day skiing, flying down slopes, Nicky had a grin plastered on his face. That was his first time skiing and he was flushed with cold and thrilling speed.

I didn't know that Patrick had found Nicky's pipe, shaped like an eight ball, in the cushions of the couch in the Orange Room, had scraped out the gunge and smoked it with his friends, tripping wildly. Gunge was the tarry black residue that accumulated in our pipes, a potent concentration of marijuana, angel dust, and hash. With a pen-knife or paper clip, you scraped the black tar that collected in the screw threads, then smeared this goo on a piece of tinfoil — it looked like hash oil, only blacker — and held a lit match underneath the foil until the gunge bubbled and smoked. With a straw or plastic pen casing, you inhaled the smoke.

I didn't know that Patrick and his friend Zach got so dusted once that, with hallucinatory, superhuman strength, they pushed the foot-ball sled at the junior high all the way down the field, across the gravel track, and into a swamp. What type of role model had I been for Pat-rick, smoking dust on that ski trip, Patrick walking into the room that still reeked, leaving drug paraphernalia in the Orange Room?


One night a few weeks after Nicky and I started going out, we were down cellar in the Orange Room, which was always cool and damp like a cave. We locked the door with the eye hook, and on the narrow couch we explored each other's bodies like little kids turning over rocks in the woods, Nicky asking if he could see what I looked like *down there,* the moment tender, with the innocence of children playing doctor, our exploring healthy because we were falling in love. We exchanged Christmas gifts that night, a brown flannel shirt for him, which would wear out, and for me a gift that wouldn't fade, Neil Young's *After the Gold Rush.* Nicky watched for my reaction as I placed the record on the turntable, lowered the needle. Music brought out a serious, soulful side of him. He was teaching himself to play guitar; a poster of his idol, Jimi Hendrix, hung on his bedroom wall.

You can only hear Neil Young for the first time once, like your first time getting high, or sex for the first time, your hand down a boy's pants, the heat and surprising metamorphosis, flesh filling your palm, silky and smooth. Senses dilated, you stand on a threshold (or precipice), and the world beckons, everything new for the last time. Neil Young beckoned with his strained raw falsetto: *I am lonely but you can free me / all in the way that you smile.* That album was the first I owned, not borrowed from my older sisters, and I attended to it as if it were poetry. Nicky inspired in me the habit of listening, of getting out of my head for a change. *It's only castles burning / find someone who's turning / and you will come around.* I was turning, like the wooden bowl I lathed in shop class in ninth grade, spinning into shape but shapeless still, not yet come around. Often I sat in the Orange Room by myself, lifting the needle after the last track on that album and setting it on the first again and again: *When you were young and on your own / how did it feel to be alone?*

If there were guidance counselors at Walpole High School who might have noticed a kid sinking, I never saw them. But there were over a thousand students in the school, and in those peak-drug-use years too many kids were sinking. A couple of teachers tried to help me. Coach Brainard asked Sue to convince me to try out for the basketball team. We played pickup games in gym, so Brainard must have seen that I could play. And Miss West, my algebra teacher, a soft-spoken, pale-faced woman who wore ballet flats, said to Sue one day, "What's happening to your sister Maureen? She's smart. Didn't she used to do gymnastics? What can we do to help?"

One afternoon Mr. Gurkin, my Spanish teacher, pulled me out of detention and led me into an empty classroom down the hall. Detention was the only time I did homework, held captive for an hour, a tiny sentence. The punishment was academically helpful, though I didn't earn enough detentions to keep up with schoolwork. Mr. Gurkin gestured for me to sit. He asked how I was doing, and then he launched into his speech. He told me I was hanging around with the wrong kids: Nicky, Paula, Alison. "They're lowlifes," he said. I was wasting my time with them.

This was not a good opening. Paula was my best friend. I saw her every day, called her every day. Why was Mr. Gurkin picking on her?

She was quiet in our Spanish class, not a back-talker like me. And Nicky was my boyfriend. I *loved* him. How did Mr. Gurkin even know Nicky, who didn't take Spanish? Nicky was not one of the troublemakers summoned over the PA every morning to the vice principal's office — Fitzgerald, Giancarlo, Miller.

Mr. Gurkin was the cool teacher, young and long-haired, rumored to be dating a cheerleader (she was eighteen). I had him for homeroom, too, which is how he knew I came to school stoned most mornings, sliding into my seat just as the bell rang. Mr. Gurkin continued in this vein of hanging around with "losers," his voice earnest, his effort sincere, but he must have seen that he wasn't convincing me, wasn't penetrating my tough exterior. "I know you come from a broken home," he said. That was the first time I'd heard those words to describe my family: broken. That phrase always meant someone else, some family from the "inner city," poor, with an alcoholic father or a drug addict mother, parents who beat their children. "Broken home," when it was first used in the mid-nineteenth century, meant a parent had been lost to death. By the 1950s, a broken home was associated with divorce, and in the 1970s it was frequently cited as the cause of juvenile delinquency.

I didn't see my family as broken, but it was true. *Broken* was the perfect word to describe what happened to my family when my parents separated; we splintered like a mirror dropped to the floor, the whole broken into individual units. Mr. Gurkin's statement was accurate, but it felt like an accusation. It was impolite of him to notice, and it was none of his business anyway. How did *he* know? The rough edge of the consonants in that word cut into me — broke, broken down, broken home. "It's not *broken,*" I said. I was affronted that he both knew and used that private information against me, that it meant something to him that it didn't mean to me. I was broken and needed fixing; he was going to help glue me together like Humpty Dumpty. I don't remember how the conversation went after that, only that anger clotted my throat. I seethed as he lectured. At some point I said, "Fuck you," my big-girl version of "Shut up." I pushed my chair back and walked out.

There was great power in refusing the help of others — the power of believing that you didn't need help, that everything was fine, the power of believing that you were powerful enough to solve your own problems, to navigate the world, to take care of yourself. *Fight your own battles.* I resented Mr. Gurkin for trying to save me (as if I needed saving), but I resented him, too, for giving up so easily, for being put off by my anger, which was a bluff, a front for hurt and sadness.

Mr. Gurkin never tried to help me again, and instead he turned against me. In Spanish, Craig Wyatt, a football star, a favorite of Mr. Gurkin's but a mediocre student, sat behind me. One day Craig asked me for an answer during a test. Surely we would be caught with Craig's unsubtle whispering, so when Mr. Gurkin turned his back, I whisked Craig's test off his desk and gave him mine, which was already done. I filled in the answers on Craig's test and, when Mr. Gurkin was distracted, exchanged our papers again. This became our practice for every test.

One day Mr. Gurkin noticed some movement. He stared at me. "Are you cheating?" he asked.

"No," I said, but he didn't believe me, the girl from the broken home who hung around with losers, so he asked Craig, the tone of his voice no longer accusatory.

"Was she cheating?"

Craig said, "No," and Mr. Gurkin believed him, no further questions. I realized later — incredulously — that Mr. Gurkin thought *I* was cheating off Craig. At the end of the term, Craig gave me a bag of pot, thanked me for "passing" him.

Mr. Gurkin turned against me, and in turn I turned against him. He announced a dress-up day in Spanish — no jeans. On that day I showed up in dungarees, with a jean shirt, jean vest, jean jacket: a sartorial middle finger. "You could have worn something nice," he said to me and Paula, the only two who'd rebuffed his "no jeans" rule, before he threw us out of class. Partly Paula and I planned to defy his ridiculous decree, but also I didn't have many nice clothes, like the dresses and skirts the other girls wore. Mr. Gurkin knew I was from a broken home, but he didn't understand that I was broke.

•   •   •

On summer days when Sally and I were in grade school, we wrote short stories, mostly about jewel thieves and bank robbers, influenced by Encyclopedia Brown, boy detective (and his helpful friend, Sally), and by our favorite TV shows, *Perry Mason*, starring the hulking Raymond Burr, and *It Takes a Thief*, with Robert Wagner, a cat burglar turned government spy. "I've heard of stealing from the government, but *for* the government?" Wagner said over the opening credits, three years before Nixon's lackeys burglarized the Democratic National Committee's office in the Watergate complex. In my stories, the main characters were heroines, never criminals, but years later, at fifteen, I plotted a burglary with my friends Paula and Alison.

I don't recall how the idea germinated, but maybe from conversations with Nicky, which I recorded in my diary, my criminal education writ small in girlish cursive.

> December 2, 1975 — Nicky and I had the most insane talk, about finding a million dollars and going to California, and how he and Vince psyched on doing B and Es, that's breaking and entering.

I stood with Nicky that Friday night outside Blackburn Hall, while inside a local band covered Led Zeppelin and Aerosmith, the muffled percussion thumping into the chilly night, Nicky's honey curls lit like a halo by the streetlamp he leaned against as I leaned into him, falling into his pillowy goose-down jacket as I kissed his warm mouth, my cold cheek against his smooth cold skin. I remember that kiss, the seal of our mouths, a hollowness like a tunnel between us, waiting for his tongue.

One night Nicky was supposed to come to my house but never showed. He called near midnight, said he'd been arrested for breaking and entering. Somehow in the course of the night, he'd broken a finger. He was absent from school for a few days, but he delivered an envelope to my house, had his mother drive him over and dropped it in my mailbox. The card was a photo encased in Mylar of a rising — or setting — sun above a mountain lake, with a quotation from Thoreau: "If a man does not keep pace with his companions, perhaps it is because he hears a different drummer. Let him step to the music which he hears, however measured or far away." I can see Nicky in the stationery store with his mother — her sculpted cheekbones and slightly worn beauty, intense green-brown eyes, which Nicky had inherited — picking through cards until he found the perfect one, the ripple of sunlight on water and Thoreau's quote as if written for us, the outsiders we felt we were. Nicky apologized for his messy four-fingered cursive and signed with an X and an O.

His arrest was listed in the police log of the *Walpole Times*, which I clipped and pasted into my diary, proud of him, this robber boy, like the Wild Colonial Boy, like the Artful Dodger, reckless outlaw boys I was drawn to. Obedience to rules seemed cowardly then, rebellion and daring admirable. Sally wore a button on her dungaree jacket that I coveted: QUESTION AUTHORITY.

When I saw Nicky next, he gave me an amber prescription bottle filled with coins that he'd stolen during an earlier B&E. He asked me to hold the bottle, perhaps so he wouldn't be caught possessing stolen goods. The prescription was dated 9/16/75, from a pharmacy in Wrentham, filled not with Motrin, as the label said, but with coins, some foreign, a few worn buffalo nickels, wheat pennies.

That summer after our sophomore year, as part of his probation Nicky worked on a grounds crew, spreading cedar chips on nature trails in Walpole, a Comprehensive Employment and Training Act (CETA) job, the result of federal legislation passed in 1973 to provide public service work to low-income people, youth offenders, and newly released prisoners. Nicky was fortunate in that a couple of years before he was arrested, Jerome Miller, a sociologist and the new director of Massachusetts's Department of Youth Services, closed all but one of the state's youth correctional facilities, an act that the nonprofit Center on Criminal and Juvenile Justice called "the most sweeping juvenile justice reform in history." Prior to that, the average age of a boy sent to a state detention center had been fifteen and a half, Nicky's age, and the most common crime he'd committed was breaking and entering, Nicky's crime.

Miller made unannounced visits to the state's youth detention centers and documented children being beaten and raped or sent to solitary confinement, facilities in squalor. One day in January 1972, Miller staged a breakout at the Lyman School, the nation's first reform school for boys and the most notorious in Massachusetts. Albert DeSalvo, the Boston Strangler, had spent time there. Lyman was the last of the state's youth detention centers that Miller would close, and on that January day Miller led a one-hundred-car caravan of social workers and reporters to the school, where he and the Lyman boys, wielding sledgehammers, smashed the locks on secure cells. Then Miller and his crew drove the boys to the University of Massachusetts in Amherst and lodged them in dorms while his staff found alternative community-based programs for all of them.

After Miller closed the reform schools, kids who broke laws, abused drugs, or ran away were evaluated by pediatricians, psychiatrists, social workers, and teachers, then sent to reputable private facilities like Longview Farm in Walpole or to small group homes, forestry camps, private prep schools, Outward Bound, specialized foster care, art schools, military schools, boarding schools, community drug programs, or "wherever else a program might develop in response to the needs of a troubled or troubling teenager." Many kids, like Nicky, were sent home with oversight and a job.

Miller's plan worked. Massachusetts, which had the eighth largest population of kids in detention centers in the early 1970s, reduced the daily population of its largest locked detention center from 250 to 25 with no increased risk to the community. Decades after Miller shuttered reform schools, Massachusetts's one-year recidivism rate for juvenile offenders was less than 25 percent, far below that of states that still jail youths, like Wisconsin's 70 percent one-year recidivism rate and Arkansas's 60 percent three-year recidivism rate. Massachusetts's confinement rate of juveniles is still among the lowest in the nation.

Thanks to Jerome Miller, instead of going to reform school Nicky worked as a landscaper. Nicky was fortunate, too, because he was white and from the suburbs. Kids commit crime equally across racial and class lines, but even wealth doesn't defeat racial discrimination. Poor white kids are less likely to go to prison than rich black kids. One study found that only 2.7 percent of the poorest white children (the lowest tenth of income distribution) went to prison, while about 10 percent of affluent black and Hispanic youths did. Once kids are caught in the criminal justice system and labeled — psychopath, sociopath, unsocialized, aggressive — people continue to see them that way. Worse, that's how they begin to see themselves.

I never thought of myself as a juvenile delinquent, probably because the cultural image of a delinquent didn't resemble me, a white girl from the suburbs. I thought I was an average teenager in the mid-1970s, taking drugs, stealing, skipping school, involved in petty crime. It seemed normal, but girls were sent to reform school for infractions far less serious than what my friends and I had done.

It's possible that Nicky's B&Es inspired mine. I didn't just want to be with reckless outlaw boys or listen passively to their escapades. I didn't want to be a cheerleader on the sidelines as I had been the year before. This was the second wave of feminism, a term I didn't know but absorbed through osmosis. Paula and Alison and I could get wasted just as well as the boys. We could skip school and hitchhike and have sex and swear. Alison spit — this striking girl who drew attention from men and boys whether she wanted it or not — could hawk a loogie as well as any boy, her act the antithesis of femininity, a rebuke, perhaps, of men ogling her. She'd clear her throat to gather phlegm, forming her pretty mouth as if she were going to whistle, her chest heaving, then *thwoop* and the triumphant splat on the sidewalk. Alison ritualized this act she'd appropriated from boys, got a kick from their surprised or disgusted reactions. I envied her ability and her nerve.

For my part, in Miss Geoghegan's social studies class, "America in the Twentieth Century," I argued that girls could do anything boys could do. "ANYTHING!" I screamed at David Simpson across the room after he smugly touted the superiority of boys. I despised his proprietary sense of privilege, how he assumed the power to limit what I — or any girl — could do in the world. I raged. Miss Geoghegan smiled, her shiny, bouncy brunette hair always slightly mussed, and tried to calm me with her hand on my shoulder.

After I was no longer babysitting, after I quit my weekend job selling hot dogs at the stadium, I had no money for drugs. At the mall, Paula and I sold raffle tickets we'd stolen from a school fundraiser and used the money for angel dust. Paula stole scales from the science lab. While our classmates put away the Bunsen burners and washed test tubes, she'd tuck a scale under her snorkel jacket. I acted as the lookout. In the hubbub of the cleanup, the bell ringing, everyone gathering their stuff to leave, nobody noticed Paula carrying a bulky object under her coat and walking out the back door of the classroom. She sold the scales to her brother, Duane, for his angel dust operation. After she stole the fourth one, Mr. Hood, the chemistry teacher, must have noticed, because the scales were then tagged and tracked, stored in a locked cabinet.

One night Paula and I made an interesting discovery when her mother sent her to Barnes's corner store for a carton of cigarettes. Each night from six to seven, Mrs. Barnes worked at the store so that her husband could walk to his house next door and eat dinner. Mrs. Barnes was tall and thick but seemed frail, with hunched shoulders. She wore pink foam curlers in her pink-hued auburn hair under a hairnet and a snap-down housecoat under a sweater, even in summer. That night when Paula asked for the cigarettes, Mrs. Barnes shuffled into the curtained-off back room and yelled, "What was the brand?" In the dozen steps from the counter to the back room, she'd forgotten.

While she was searching for the cigarettes, Paula eyed the tackle box on a stool behind the counter. Mr. Barnes had stopped using a cash register after Doreen Randall's older brother, Dale, with a nylon stocking over his face, conked Mr. Barnes with a baseball bat and grabbed the cash drawer. Dale ran down Bowker Street, dollar bills floating out of the tray. That was the story, anyway, which was why we combed his getaway path for stray bills. But the newspaper tells a different version, with Mr. Barnes doing the conking, perhaps in a different robbery.

> *Walpole Times,* Dec. 4, 1975 — Clifford Barnes Sr. said two youths entered the store and demanded his money. When one of the youths pulled a knife, Mr. Barnes hit the youth in the hand with a club, and the two would-be robbers fled down Bowker Street.

While Mrs. Barnes fumbled in the back room, Paula hoisted herself onto the counter and grabbed a few bills from the tackle box. After that, every couple of weeks when Mrs. Barnes was on duty, Paula fished money from the tackle box. She grew more daring, grabbing fives and tens, though never enough to arouse suspicion, we thought. One night Paula paid for the cigarettes with a $20 bill and we watched Mrs. Barnes place the twenty in an empty plastic half-gallon ice cream tub set atop a refrigerated case, hidden behind a rack of chips.

The next time Mrs. Barnes was in the back room fetching the carton, Paula reached into the plastic bin and took a $20 bill. Like all thieves, Paula got greedy. One night she grabbed a fistful of bills from the tub — $80. I worried that the sizable amount would be noticed. I thought she'd gone too far, even though I was happy to share the drugs she bought with the money. We waited a few weeks before entering Barnes's again, but when we did, we encountered the old man at the dinner hour. Mr. Barnes glared at us. "You weren't expecting to see me, were you?" We shrugged as if we didn't know what he was talking about. "Get out of my store," he said. "And don't ever come in here again."

Outside I burned with shame. I hadn't stolen the money with my own hands, but I was an accomplice, there for immoral support. We justified stealing from Mr. Barnes because he was a "perv" who used to grab our sweaty pudgy hands when we passed him money for penny candy and held those hands for an uncomfortably long time, the price girls paid for a sack of Mint Juleps and Red Hot Dollars and Atomic Fireballs. Worse, when my mother sent me to the store for baking soda or condensed milk, shelved in the dark corner behind the ice cream freezer, Mr. Barnes would come around the counter to assist me. He'd put his arm around my waist, crawl his fingers across my back if I was wearing my two-piece bathing suit because I'd been playing in the sprinkler before my mother had sent me to Barnes's for the thing she needed.

Still, I felt horrible. I'd been going into that store since I was a girl in a bathing suit on a bike buying Bonomo Turkish Taffy or freeze pops. Mr. Barnes had watched me grow up, as he had all the kids in the neighborhood, but now he hated me. I suppose he could have told our parents, but he never did, because none of them ever said anything. He could have called the police, but he didn't do that either. Perhaps Mr. Barnes was more hurt than angry. He seemed disappointed, in me particularly. I sensed that he didn't like Paula, who was tainted by the reputation of her brother, who'd been a star athlete in high school but years later was known for trouble.

One day on the way to school Paula was near tears, worried that her brother might be sent to jail after his trial that morning on a charge of possessing stolen property, valuables from a house that had been burglarized. But when we walked into Paula's kitchen after school we found Duane, still in his suit coat, celebrating his acquittal. The police had found the stolen items in his family's garage, but Duane had been saved by the Fourth Amendment, which bars illegal search and seizure; the search warrant had not listed the garage in the description of searchable areas.

The B&E might have been Alison's idea — the house we planned to rob was across the street from hers. I could imagine Alison looking out her bedroom window, thinking about how the family went skiing, the mother and father and their three children, whom Alison used to baby-sit, how they left behind two spinster aunts who were half deaf, how they kept money in their refrigerator.

Paula and Alison and I spent weeks rehearsing the B&E, our burglary premeditated to an absurd degree. Planning the B&E felt like writing a short story about cunning thieves with a clever plot, every angle studied, the escape route mapped, though never once did we think about the consequences, about the possibility of getting caught, about the two old women in the house. If circumstances were different, we might have been planning a play we wanted to stage, with costumes and set designs. We might have channeled our rebellious energy toward political protest, environmental action. I'd participated in the first-ever Earth Day in 1970, when I was ten, picking up litter along a stretch of road, and in seventh grade, for my science report on "any animal," I'd chosen "man," the only species that pollutes its habitat. I shot Polaroids of the car dump in the woods behind my house, and my father drove me into Boston to photograph pollution billowing from factory smokestacks. Instead, at fifteen we put our heads together to plot a crime.

In spite of our elaborate discussions, our plan was simple. We reviewed it again and again, maybe to convince ourselves we'd go through with it. We would do it when the family was away on vacation. The back of the house abutted woods, so we'd approach and escape through the backyard after the aunts were in bed. We'd sleep at Paula's house so Alison wouldn't have to go home afterward; she'd have an alibi. Because we spent so much time planning our B&E, we were compelled to follow through; the momentum, the commitment we spoke aloud — nobody wanted to be the one to chicken out.

That Friday night, drinking cans of beer Paula snuck from her refrigerator, she and I took a shortcut through the woods, cat-footing on a plank across a stream, emerging at the train trestle behind Friendly's plaza, where we waited for Alison underneath the overpass. I lit a cigarette, fetishizing the ritual — tapping the pack against my palm to compress the tobacco, pulling the cellophane tab, crumpling the silver foil, striking a match, then the pleasing scratch and burst of flame, the whiff of sulfur sharp in my nose, the heat on my face as I sucked in smoke, snapping my jaw to blow smoke rings.

I wasn't thinking of what we were about to do — break into someone's house — or maybe it still didn't feel quite real, as if something would derail us yet, that we'd play this out until the last minute, a scheme that would never be enacted, even though we'd set it in motion by meeting that night on the tracks, sitting on the cold metal rails, blowing smoke rings with my hot cigarette breath, circles hovering in the frigid air as if paused in time.

One day when I was eight, two police cars zoomed down our street and screeched to a halt between our house and the Wagners'. The policemen leapt out of their cars and strode across the lawn toward the woods, their cruiser doors ajar, radios emitting tinny voices and scratchy static, each cop touching the handle of his gun holstered at his hip, ready to draw, like that cartoon character, Quick Draw McGraw. The officers disappeared into the woods where my brother had a fort, where we skated on the swamp behind the Richardses' when it froze, past the dump with a half-dozen bullet-pinged car bodies, where I hunted for green and red plastic shotgun shells.

Neighbors gathered on the sidewalk, quietly chatting, but after twenty minutes or so the cops emerged empty-handed, jumped in their cruisers, and zoomed back up the street. Whomever they were hunting behind our house, he was still out there.

Walpole Prison was called "escape-proof," but shortly after it opened an inmate named John Martin escaped by tucking himself between the body and frame of a milk truck, *Cape Fear*–style. Unfortunately for Martin, cops spotted him two hours later walking down Main Street in the bordering town of Norfolk. In Martin's second attempt, he wore a guard uniform he'd stolen, dyed his skin with tea bags, drew a crayon mustache on his face, and then brazenly strolled toward the exit, nearly making it before a tower guard grew suspicious.

That same year Frank Drozdowski, an ex-Marine serving life for murder, walked away from a coal silo outside the prison where he worked as a trusty. He tramped through the woods and cedar swamp surrounding the prison for five hours, then stumbled onto the road near Pondville Hospital just three hundred feet away, where he was caught. Another inmate, Michael Thompson, wedged himself between some furniture on a delivery truck and passed through Walpole Prison's gate, but the truck was traveling only a mile down the road to Norfolk Prison. When Thompson stepped out of his hiding place, he was still behind prison walls.

After the prison opened in 1956, the town of Walpole's selectmen asked the commissioner of corrections to sound an alarm when a prisoner escaped, and so occasionally we heard sirens, deep resonant blasts like tornado warnings in the Midwest. When the alarm sounded, how were we supposed to react? Lock ourselves inside? Form a vigilante posse? Sue was afraid that escaped prisoners would hide in the woods behind our house, but my mother assured her that escapees would want to get out of Walpole as fast as they could. Still, they'd have to pass *through* Walpole.


Maybe it was Robert Dellelo the cops were looking for that day they searched the woods behind my house. After he escaped from Norfolk Prison on September 21, 1968, there was a massive manhunt around Walpole and Norfolk, involving 125 local and state police, bloodhounds, a helicopter. Six years earlier Dellelo and another man had attempted to rob a jewelry store in Boston but tripped an alarm. Outside the store, after the two ran in opposite directions, Dellelo's partner shot and killed a cop. Even though Dellelo was a block away, the killing was considered part of the commission of the crime, and at twenty-one he was convicted of murder and sentenced to life without parole.

Dellelo served a few years in Walpole Prison, then was transferred to medium-security Norfolk for good behavior, and there he methodically planned his escape. He studied the twelve strands of electric wire on the first barrier, a chain-link fence, and saw how he could short-circuit the fence and scurry beneath it. He observed the radar system and noticed a gap in its range of detection. From the paint shop he stole a thirty-foot rubber cord, to which he attached a grappling hook fashioned from bucket handles. Using skills he'd learned in Walpole Prison, he removed the lock assembly from a hallway door, then molded a key from aluminum and reinstalled the lock before anyone noticed it was missing. He picked the locks of the supply room and stole wire cutters.

Dellelo patiently waited for a forecast of fog. The night of his escape, he placed on his pillow a plaster-of-paris dummy of his head with broom-bristle hair and draped a fellow inmate's dirty clothes on a chair in his cell to confuse the bloodhounds he knew would track him. (When the dogs tried to follow the scent from the clothes, they ran in circles inside the jail.) He dropped ten feet out a window, but contrary to newspaper reports that blamed the fog for the fact that Dellelo "vanished without a trace," the fog arrived long after his escape.

Dellelo dodged the radar, then ran unseen across a 150-yard no-man's-land to the electrified chain-link fence. Protected by a rubber pad he'd stolen, he snipped the fence and crawled underneath. He bolted another fifty yards under floodlights and reached the twenty-foot concrete outer wall, its top garlanded with electrified barbed wire. He scaled the wall with the rubber cord, and at the top, wearing squirrel-fur gloves lined with rubber and boots he'd rubberized, he cut the electric wire, then slid down the other side. He dashed to the tree line, where he covered himself in pepper to mask his scent.

And then he ran.


Dellelo escaped along the train tracks through Walpole, ground I knew well. As the crow flies, it's about a quarter mile from Norfolk Prison to the railroad tracks behind St. Jude's Church. On summer days in junior high, my friends and I rode our bikes past the prison and cut across the tracks to a kettle-hole pond, where we swam and gossiped. In eighth grade I rode along these tracks on the backs of minibikes that were more like small motorcycles, holding tight to boys I liked, like Christopher Nash, with his deep brown eyes, shoulder-length chestnut hair, a sexy silver cap on his front tooth, a rooster tail of gravel spitting off the back tire as we sped along. I hugged his waist, pressed my cheek into his back, inhaling the musky scent of his leather jacket, until he pulled over in some woods and we lay down on leaves and he pushed his tongue into my mouth.

One night in eighth grade, Alison and I and another girl, Jackie Conley, walked those tracks to Dante's, a six-lane bowling alley in the basement of a restaurant, where we set fire to paper towels in the girls' room and clogged the sink dousing the flames. The boys — Adrian, Carl, and Ricky Strickland, my boyfriend in eighth grade (who later landed in Walpole Prison) — whipped the candlepin balls so hard they skipped wildly across lanes, the noise drowned out by the jukebox upstairs. Jackie found a carton of macaroni in the supply room and scattered it all over the stairs so that it would crunch beneath the feet of the manager whenever he came down. We left quickly after that, walking along the tracks to stay hidden.

Near the town center, Dellelo would have encountered an intersection of four rail lines behind the Kendall Company, Fiber Products Division, which I passed when I walked or rode my bike downtown, always a hum emanating from the building and all around the factory white fluff in the air, like after you blow on a dandelion that's gone to seed. The tracks ran behind the row of stores one block deep on Main Street — Holt's Clothing (where my mother stole clothes), First Sandwich Shop, Tee-T's restaurant, where my father used to take us for fried clams and pizza.

At that junction one set of tracks ran slightly north, past the end of my dead-end street, where my brother Patrick and his friends played chicken, dashing across the tracks as trains bore down, where one day they broke into a stalled railcar and found — like a fantasy they'd conjured — cases and cases of Miller Beer, the sixteen-ounce cans. They hid as many cases as they could in the woods, and then someone's older brother came in his truck and hauled more.

As a kid I loved the train tracks, which seemed both forlorn and hopeful. When I was a girl, Sally and I and our friends packed lunches and rode our bikes to the dead end, picnicking on the tracks, the wooden ties smelling pleasantly of creosote, of tar, eating bologna sandwiches as we waited to feel the earth vibrate beneath our feet, and then the rumble and the sudden deafening roar as the train thundered by, the *clickety-clack* of each car passing over the joints, afterward the air singed and changed. I'd squat by the creek that gurgled in a ditch alongside the tracks, searching for translucent crayfish. They were nearly invisible, but if I sat perfectly still, I'd spot them scuttling along the mud bottom, leaving a faint trail like an echo.

The swamp on the far side of the tracks — you had to cross a log laid over the creek — was edged with blueberry bushes drooping with fruit. Sally and I spent hours picking berries on days so hot the air shimmered off the iron rails, hot enough to scorch your fingers. My mother turned the kitchen over to us and we'd roll out dough, stir the filling, and shove the pie in the oven, then sprawl under the shade of the oak trees in our side yard as the pie baked, happily staring at the sky.

I liked being in that forbidden place, hushed but for the buzz of insects, birds thrashing in bushes crowding the tracks. Sometimes we'd see a figure coming toward us, a man walking the tracks alone in broad daylight, T-shirt slung over his shoulder from the heat, and we'd hide and wait for him to pass. *Who was he? Where was he going?* I'd stare down the tracks until they converged to a single point and disappeared, like a magic trick.

When I was sixteen, old enough to feel the tug of the world, after Sue had moved out for college, I had my own bedroom for the first time, with a double bed and a lime-green bedspread. My head was just below a window that faced the dead end, and lying in bed on balmy summer nights I'd watch the curtain lift and billow on the breeze as if it had desire. I'd hear the chuff of the train near midnight and the faint whistle and feel already far away.

At that juncture downtown, if Dellelo had turned slightly north onto the tracks that ran past the end of my street, he'd have eventually reversed direction as the line curved south and continued to Rhode Island. Instead he ran straight, chugging along the tracks that ran northeast toward Boston, passing behind the A&P plaza and Friendly's, where Paula and I waited for Alison on that cold February night, Dellelo a young man, just twenty-seven, a prisoner breaking out; my friends and I just fifteen, fledgling criminals, breaking in.

After Alison arrived, we walked along the tracks and then cut into town. We took the shortcut through the field behind United Church, where Davey Winters had hanged himself from the wrought iron fire escape the year before but where everyone still went to get stoned, where I'd eaten the Popsicle I stole from Mimi's Variety, where in a few months Alison would try to kill herself by swallowing a bottle of pills she'd steal from Betro Pharmacy. We stopped at the edge of the woods behind the house we were going to rob. I could see across the street into Alison's living room, the silhouette of her mother in the window against the flickering gray light from the television.

We skulked across the lawn to the back door, which was unlocked, as if the family were expecting us, and we walked in. The house was dark but for patches of street light filtering through the windows. The aunts were upstairs asleep, we assumed, so we had to be quiet, even though they were half deaf. I started to have a nervous laughing fit, probably from the tension of suppressing my fear and gnawing guilt. I knew what I was doing was wrong, but I was doing it anyway.

I saw Alison lifting silverware from the drawer of a sideboard and stuffing the pockets of her coat. For all our elaborate planning, we'd forgotten to bring a bag. At the liquor cabinet, Paula shoved a quart bottle of something in her sleeve. Alison had mentioned that the people hid money in the refrigerator, so I looked in there. In back behind the milk I saw a small tin bank in the shape of the world, Africa a yellow blotch. As I reached for the globe bank I knocked an egg out of its preformed manger in the refrigerator door and it smacked quietly on the linoleum. Paula and Alison didn't even hear it.

Sometimes I wonder, did this really happen, the egg breaking? I remember feeling clumsy in my thick parka, logy from beer and nervousness, seeing the eggs in the refrigerator under the bluish fluorescent light, the sensation of something slipping. How absurd the moment was, the invasion of someone's refrigerator more invasive than the invasion of their home. I know I was in their refrigerator. I remember the light, the cool air on my face, the eggs, the milk, the globe bank, a sound from upstairs, someone roused, and then the three of us hurrying out the back, the screen door slamming behind us, racing through the woods and reconnoitering breathlessly in an alley downtown, the giddy anxiety in my stomach, the two-mile walk back to Paula's house. I remember prying the rubber stopper out of the globe bank, the small sum contained in that miniature world, $28.

I'm haunted by another memory that I'm not sure is real — that the aunts woke, that after we heard rustling upstairs, someone called, "Who's there?" How awful it must have been for the women to suspect someone was in their house, some dangerous person or people. I've lived alone often enough to understand how horrifying it would be to sense a menacing stranger in my house, how my heart would throb sickeningly, how I'd lose much more than whatever the intruder could take — how I'd lose my sense of safety in the haven of my home, my sense of safety in the world.

After our robbery we perused the police log, and one day there it was in black and white, almost astonishing, the name of the street, the fact of our theft, the news that we'd made, something worthy of reporting, ours one of four burglaries that night. But somehow the whole episode was anticlimactic, even embarrassing. The robbery was so amateurish, so ridiculous and shameful, that we didn't talk about it afterward.

The neighbors suspected Alison and mentioned it to Alison's parents, who interrogated her. Perhaps someone in the family remembered telling Alison about the money in the refrigerator. Otherwise how would the thieves know to look there, behind the milk and juice? But Alison had an alibi — she was sleeping at Paula's house across town. Our paltry haul amounted to less than $30, some worthless silverware that we had no idea how to pawn and so remained in a bag in Alison's closet, and a quart of Four Roses whiskey — I remember the red roses on the label.

· · ·

Ours was a "third-rate burglary," to use the words of Ron Ziegler, President Nixon's press aide. In Nixon's B&E, four of the five burglars stayed at the Watergate hotel. At least we'd had the sense to leave the area. A moment of sloppiness led to their capture — instead of placing tape over the door locks vertically, they taped them horizontally, so that the tape was visible on the front of the door and was noticed by an alert security guard.

In ninth grade on Fridays in Mr. Klein's civics class we played College Bowl, a trivia contest of not-so-trivial current events: the oil embargo, Vietnam, Watergate, Nixon, who'd resigned a month before school began that year. We knew some facts and names — just enough to win a candy bar — but we didn't delve into the issues. (It would have been edifying if we'd taken a class trip to Norfolk Prison that spring of 1975, when inmates staged a reenactment of the Watergate crimes for an audience that included Senator Ted Kennedy.)

From what I'd gathered about Watergate, it was Nixon himself slinking around in black clothes and, I imagined, a nylon stocking over his face. But what was he trying to steal? That, I didn't know. I didn't know that his henchmen were not taking something but leaving something, bugs (or repositioning the bugs they'd planted earlier, because they weren't recording properly). But Nixon *was* taking something — secrets.

Was the president's burglary an imprimatur for us to do the same? Was the rise in crime in the 1970s, ours included, attributable to a trickle-down effect? "It is a classic idea that a whole community may be infected by the sickness of its leadership, by a failure of ideals at the top," Anthony Lewis wrote in the *New York Times* in 1974. "We are infected by corruption at the top." As the nation goes, so goes its towns, its families, its citizens. Whatever reasons allowed me to forget my moral sensibility, pounded into me by my mother using Walpole Prison as stark example, whatever reasons placed me in a strangers' house at night taking their things, in the end our B&E was more successful than the president's because we were never caught, in spite of the pathetic evidence I'd left behind: a mitten.

When I think of the B&E now, I'm glad I dropped the mitten, this clue that led the family to suspect Alison, because maybe then the two aunts were not afraid in their home after our burglary. Maybe they would take comfort knowing that the break-in was not some bad guy, some dangerous escaped prisoner, just some sad troubled girls.

6

*Hello World*

IN OUR JUNIOR YEAR PAULA BEGAN TO DEAL ANGEL DUST FOR Duane, which meant that we had a seemingly endless free supply of dust. At Fernandes, Duane or Wayne Kosinski or Rod Tyler passed through Sue's checkout lane with jars of parsley flakes, which they soaked or sprayed with liquid PCP. Maybe they picked Sue's lane because Jeff, her boyfriend, was dealing for them. But even if a store manager noticed, it wasn't illegal to stock up on dried parsley.

Angel dust was fairly easy to produce and enormously profitable. A $100 investment in the precursor chemicals could yield $100,000 worth of angel dust on the street. The main ingredient, piperidine hydrochloride, was available for purchase without identification, sold by companies that supplied university and research labs. Around Los Angeles in 1977, police were destroying PCP labs at the rate of about one per week. Smaller home labs were run by amateurs mixing PCP in pots and pans in their kitchens, which was, according to the Boston Police Drug Control Unit, "a terribly easy thing to do." In the early 1980s in South Boston, two brothers were producing $15,000 of angel dust weekly out of their mother's apartment before they were busted.

We heard occasionally that there was a bad batch of dust, to watch out because someone had made angel dust with Raid or embalming fluid. There was a rumor that formaldehyde was stolen from the high school biology lab, from those gallon glass jars of clear liquid with pale pink pig fetuses floating inside. One night outside a Fleetwood Mac concert at the Patriots' stadium in Foxborough, where I'd sold hot dogs when I was fourteen, Nicky's sister, Andrea, and I bought a joint of dust from a stranger for $2. The joint smelled rancid — not the usual stale dried-herb smell — but we smoked it anyway.

We joined a knot of kids outside the chain-link fence in a poorly lit corner of the stadium, where some guy with wire cutters was patiently clipping a hole link by link. Thirty or forty people shook the fence until it collapsed and kids spilled onto the concrete promenade inside, tumbling over each other and running off. The last stragglers — me included — were sprayed with tear gas by the security guards. For hours I was stuck on the stadium lawn in a grip of bodies, my head pounding from smoking that greasy joint, my ears ringing from the too-loud concert, my eyes boiling from tear gas.

Duane gave Paula a stack of small square manila envelopes, big enough to fit a silver dollar, filled with a gram of angel dust that he'd carefully weighed on the four-beam scales Paula had stolen from the chemistry lab. As compensation for her distributing dust to kids in school, Duane gave Paula free grams. We smoked dust every day, often twice. When we ran out of the free grams from Duane, we'd buy a gram, if either of us had money, or we pinched from each envelope, not enough for anyone to notice that the packet was light but enough to get us dusted.

When Paula started dealing, she had instant cachet. People sought her, needed her, or needed what she had, which seemed like needing her. She seemed to embody her new stature: a certain looseness to her gait, her neck held a bit straighter, shoulders back; she looked taller. She walked with her hips thrust forward, wearing Frye boots and a leather coat, dangly earrings, her thick reddish brown hair rippling down her back. She developed a slight slur to her speech, s sliding into z, as if it were too much effort to articulate.

There was a party in someone's house on the west side of town, near the prison. I don't remember how I got there, or with whom, only that

I was wasted on angel dust and alcohol, a lethal mix — my mind erased and my body incapacitated. I could barely stand and I was seeing double. The kitchen was packed, people pressing around me, closing in on me. *I need air.* I had just enough animal sense to stumble outside, puke on the lawn. The rest of the night was blacked out, at least until the next day when I was walking downtown. An older boy I passed on the sidewalk grinned at me. "Do you know where you live?" I expected a punch line, but there wasn't one. I walked away puzzled, but ruminating on his odd comment triggered a memory from the previous night.

After I staggered out of the house, I crawled into the backseat of some car and passed out. The car belonged to Chucky Hickman; his passenger was the kid who'd made the remark. They cruised around without knowing I was in back until they heard a moan. Chucky Hickman drove me to my house and woke me. In a stupor I opened the car door and lurched across the lawn to the house, except I was walking toward the Gibsons' house across the street. I was nearly to the front door of the Gibsons' when my mother saw me and called out, before someone — maybe Chucky Hickman — chased after me and turned me around. *I'm wasted and I can't find my way home,* Eric Clapton sang on his album released the year before. Was that the night that my mother dragged me by the collar to the bathroom mirror? "Look at yourself. Just look at yourself," she said. "I can't see," I replied, though I could see my blurry reflection, my stupid smirk.

One night when I was hitchhiking alone, dusted and drunk again, that annihilating combination, a young man in a pickup truck pulled over. When I struggled onto the running board and into the passenger seat, he said, "You shouldn't be hitchhiking." I didn't know him, but he was older, maybe in his twenties. "I don't have any other way to get home," I slurred. I remember I could barely talk. "It's dangerous," he said, pulling away. "You could get picked up by some crazy."

What if instead of being a Good Samaritan, the man who picked me up had been "some crazy," someone like Tony Costa, a clean-cut, good-looking man in his early twenties who resembled a killer as much as any boy I knew, which was not at all. Costa had worked as a carpenter in Walpole for Starline Structures, a home-improvement company, the same year he killed three young women. The next year on Cape Cod he murdered, then raped, then dismembered four other young women. Tony Costa was already dead that night I hitched a ride; he had hanged himself in his cell in Walpole Prison the year before.

What if the man who picked me up had been just an ordinary guy who saw an opportunity to take advantage of a girl so out of it that she'd never be able to defend herself, never be able to identify him? Nobody saw him pick me up at that desolate hour on the lonely streets of downtown Walpole. Nobody knew where I was. My friends could have stated only where they'd left me. *She was last seen at the corner of East and Main.*


Sometimes I wonder how I was not raped or killed or both, why was I not brain-damaged, ruined, sent away, locked up. I could have disappeared like some girls did, runaways, or girls abducted and never found, or found in a ditch somewhere. Disappeared girls were the ones who sought adventure, excitement, who rode on the back of motorcycles, like Dawn Shaheen, Miss Walpole of 1974. She represented all of us then, the aspirations of girls to be beautiful, to wear that crown. I remember the front-page article in the *Walpole Times,* Dawn Shaheen, winner of the Miss Walpole pageant for the town's two-hundred-fiftieth anniversary. I was fourteen when I stared at her beautiful smiling face in the black-and-white photo, her big doe eyes. I didn't know her, but she belonged to all of us, the whole town, Miss *Walpole* — she was *our* beautiful beauty queen. She'd just graduated from high school, had been Miss Walpole for two months, when one day she rode on the back of some boy's motorcycle and when they crashed she lost her life. Everyone said her face was unmarked, not a scratch on it; they said this in a hushed way, as if this proved something, that some force of fate had spared her lovely face.

The next year, 1975, Karen Quinlan's name was on everyone's

tongue, this girl from New Jersey in a coma from drinking too much, mixing alcohol with pills, her story the cautionary tale — "You'll end up like Karen Quinlan," forever asleep, like Sleeping Beauty. Disappeared girls were the ones who smoked and drank, who took drugs, who opened the door to strangers, who got in strangers' cars. Careless girls, wild reckless girls, girls who courted thrills, girls like me, a "girl by the side of the road," like in *Diary of a Teenage Hitchhiker,* a 1970s TV movie about a girl who got into the wrong car. The ad for the movie asked, "Where is your daughter tonight?"

The bogeyman of my childhood was a kidnapper, an escaped prisoner, the Boston Strangler. *Never take a ride from a stranger.* I knew the warnings, but I failed to heed them — too lazy, too careless, too naive, too arrogant, considering myself immune from those abstract perils. Too young. One day in tenth grade I missed the bus after school and was walking home when a man in a two-toned sedan, rusty in spots, pulled alongside and asked for directions, then asked if I wanted a ride. I was the perfect target, in my burgundy corduroy skirt and knee socks. I knew I shouldn't accept, but the man — froglike, balding, with bulging eyes, a potbelly — didn't look dangerous the way Charles Manson or the Boston Strangler looked deranged and menacing, with wild eyes. This man was out of shape, toady, like Mr. Klein, my ninth-grade civics teacher, a short, thick man with bristly hair who kept candy in his desk at West Junior High, which he gave to girls when we visited him after school. Or Mr. Hood, the chemistry teacher, with his oversized head like Fred Flintstone. Mr. Hood let me do anything I wanted. Get a bathroom pass? No problem. I knew he favored me, but I didn't know why. These men seemed harmless, these paunchy, baggy-eyed, middle-aged men.

I was tired and hot and still had two miles to walk, including up long, steep Pemberton Hill. "Okay," I said, and slid onto the vinyl front seat. The man, who wore a suit coat and shirt but no tie, chatted as he drove away. *How do you like school? What's your favorite subject? Do you have a boyfriend?* We crested the rise near the Boston Edison station, near the turn that brought me to the final mile of my journey. His question about a boyfriend was a segue to prurient interest. *Do you have sex? What do you do with him?* He saw that I was put off, alerted, and so he spoke faster, his questions increasingly graphic. *Do you give him blow jobs?*

At the bottom of Pemberton Hill, I said, "This is fine. I'll get out here." He said, "I can take you all the way home." I said firmly — or with alarm — "Let me out right here." He pulled over and I got out, and I didn't say thank you for the ride. That was my response to his lewdness — I was impolite. I walked the last mile home, Pemberton Hill like penance for my stupidity. I felt relieved, then angry, then embarrassed and ashamed, finally frightened. *How could I have been so stupid?*

How could I have been so stupid *again?* One night Nicky's sister, Andrea, and I hitched to the Flats in Norwood looking for a boy she liked. The Flats were the working-class section of Norwood, dense with flat-roofed triple-decker houses jammed close. There was a feeling of crowdedness in the Flats, like a tenement. We were picked up by a lone man in a car, someone a year or two older than us. We closed the door and he pulled away before we realized how fucked up he was. He weaved all over the road, partly because he was so drunk, but also because he wanted to kill himself, he told us.

He veered purposely into the opposite lane and we screamed. The roads weren't busy — it was a weeknight — but the few cars that swerved out of his way blasted their horns. The rest of the people in town were home, with work or school the next day, in bed or getting ready. Andrea and I begged the man to let us out, but he refused. He'd kidnapped us. Yelling seemed to encourage him, so we began to sweet-talk him, to cajole. I crouched on the floor in the backseat, braced for the collision I expected any moment.

Finally Andrea convinced him to pull over, and we jumped out in the middle of nowhere, or somewhere we didn't recognize. We had no choice but to stick out our thumbs again. This time two other Norwood boys picked us up, nice boys, and they invited us out for Chinese food, and we told them the story of the crazy suicidal kid. We ordered drinks with tiny umbrellas and the boys paid for our dinner, and they did not rape or kill us but drove us home to Walpole.

•  •  •

As a teenager I was fascinated by stories of runaways, hitchhikers, the wild girl a collective fear/fantasy in the 1970s, an archetype of the decade. First there was *Go Ask Alice,* which sold four million copies, a runaway bestseller about a runaway girl. The book was purportedly the true diary of a girl who got involved with drugs, written by Anonymous, which made the story seem more real. Alice, the diarist, died at the end, death being the punishment for girls who strayed.

*Maybe I'll Come Home in the Spring,* a made-for-TV movie, starred Sally Field, who'd played a nun before that, innocent and uncorrupted, now a druggie runaway. Even Eve Plumb, who'd played Jan on *The Brady Bunch* — my TV proxy, the middle daughter, a little whiny, a little misunderstood — became a delinquent in *Dawn: Portrait of a Teenage Runaway.* The film's theme song, "Cherry Bomb," was sung by the Runaways, an all-girl band — *Can't stay at home, can't stay at school . . . Hello world! I'm your wild girl.*

How powerfully I was drawn to these girls in pulp paperbacks and TV specials, girls who'd crossed the tracks. "Sally made spaghetti. Paula came over. Watched Sarah T., a movie about a teenage alcoholic," I wrote in my diary. *Sarah T., Portrait of a Teenage Alcoholic,* about a girl from a broken home, starred Linda Blair. It wasn't all fiction. In 1976, Mary Anissa Jones, who'd played Buffy on *Family Affair,* died of a drug overdose at eighteen. A toxicology test found high levels of PCP and other drugs in her system. Lovable freckle-faced Buffy, America's girl in the fifth-ranked show from 1967 to 1970; six years later the actress who played her dead from drugs, angel dust among them.

Why did these girls leave home? What were they looking for? I sensed there was something out there that they were running to, some ideal place, like California. Nicky and I dreamed of running away to California, that great escape, the antithesis of Massachusetts, big and sunny, not small and cold, as far away as you could get from Walpole without taking a boat across an ocean. We were not like Dorothy, who wanted to leave the glittering surreal city on the hill and get home. We wanted to find the glitter, the glamour, the grit. We wanted flying monkeys and smoking caterpillars and Big Sur and Haight-Ashbury and communes and love and everything from the sixties that we didn't know was already gone. What we yearned for was someplace bigger and more thrilling than suffocating suburbia, something compelling to which we could belong, a movement, a purpose, a point to our lives.

In ninth grade, Cathy Byrnes, a girl I knew from smoking cigarettes in the girls' room, ran away from her overbearing military father. She hitchhiked, caught a ride with a truck driver, and made it to California in three days, we heard, but then she was caught shoplifting and shipped back to Walpole. Her father sent her away to some strict academy and we never saw her again. There was something in her desperate run that I admired — the sheer audacity, the blank stupidity, the blindness to possible consequences. She seemed brave.

December 20, 1975 — Marco, Dan Valerio and Rod Tyler drove Alison and me on a couple dust runs, then had to go home. Wish I didn't have to come home at all.

Why did I wish I didn't have to come home? Where did I imagine living, changing clothes, sleeping? What would I eat? But I didn't need to run away. There was nothing oppressive about my home life; the opposite — nobody was paying much attention.

My father lived somewhere else, his visiting hours restricted. He wasn't around often enough to know what was happening. Even when my father visited, there weren't private moments between us. I remember just one day in my childhood when I had my father to myself, on my tenth birthday, when he took me to a Celtics game to see my hero, John Havlicek. On the way to Boston, we ate at a diner, which wasn't fancy — brick facade and glass-block windows — but they served my favorite food, steamed clams. After I ate an enormous mound of clams, a midden of shells left on my plate, my father asked, "Had enough?" I nodded, knowing it was piggy to want another whole plate, but I was still hungry. I wanted to sit in that restaurant eating steamed clams dipped in butter and talking to my father for the rest of my life.

One night when I was sixteen I was supposed to meet my father for dinner. I don't know how this plan came about; it was odd. Maybe he felt the need to spend time with his troubled middle daughter, or more likely my mother suggested it. That night he called me from a bar — I could hear the background noise — and said he couldn't make it. I suspected from his slurred voice that he was a little drunk, which surprised me. I'd never seen my father drunk, or even sloppy from a few drinks. He drank a beer or two sometimes on weekends, let us sip from his can when we were little, amused at our sour expressions.

On rare occasions after work my father would mix a caramel-colored drink called a highball in a short heavy glass with ice cubes clinking. The liquor cabinet held a couple of half-full bottles of alcohol, a copper shaker with the strainer top, and a weighty metal canister with a trigger you pressed to carbonate drinks, an accoutrement that we sometimes played with as if it were a toy and which my parents used only for the occasional New Year's Eve party or summer barbecue they hosted. There wasn't even enough alcohol in that cabinet to steal and water down the bottle.

My father didn't give an excuse for canceling our plans that night; there wasn't one. But I knew why he canceled. He was having too much fun at some happy hour and didn't want to drive out to Walpole to dutifully spend time with his dour teenage daughter, one who'd become a stranger to him.

"We'll do it another time," he said. I could hear the guilt in his voice.

"Okay, no problem," I said.

Just before he hung up, he mumbled, "Love you," and then I knew for sure he'd been drinking. Maybe he muttered those words hurriedly, just before the phone hit the cradle, because he didn't want to leave time for me to respond; maybe he feared I wouldn't say it back.

It was not our family ethos to declare our love for one another. My parents didn't say "I love you" as we walked out the door or hung up the phone or at night before bed, or even as words of comfort when one of us was crying or hurt. It's almost as if those words were too intimate, too private, embarrassing. My mother always reprimanded me when I said I hated anyone; *hate* is too strong a word, she'd say. Maybe *love* was, too.

My father wasn't aware of angel dust, but he wasn't entirely clueless about our drug use. When I was fifteen, he flew Sally, Joanne, Barbie, and me to California for a week; he'd amassed frequent-flyer miles working for a client there. We stayed in Culver City, east of Los Angeles, at the Stern's Motel, a mom-and-pop place, all of us in one efficiency with a rollaway cot and a foldout couch.

The first day we ate lunch at Venice Beach, which epitomized the California of my imagination, the California that Nicky and I had dreamed about — balletic roller skaters weaving along the boardwalk in white tie-up skates, the kind we rented in seventh grade at Rollerland with the rubber toe brake, skating outside in the sunshine as if they'd rolled right out of the contained loop of a rink; sun-bleached, long-haired teenage boys, tanned and beautiful, and girls in macramé bikinis or cut-off jeans; and bikers in black leather vests, chains hanging from their back pockets; and stoned-out druggies with dirty feet leaning against walls painted with psychedelic murals or trompe l'oeil. Bicycle cops in shorts laced through the crowd, kids on skateboards, buskers strumming, gay men in flamboyant clothes like I'd seen in Provincetown on Cape Cod, where my friends and I rented mopeds and buzzed along dune roads, the sheen of brown and black and bronzed skin, a stream of humanity pulsing with energy, scented with sweet musky marijuana and coconut oil.

That night Sally and I asked my father if we could walk to Venice Beach, which was two miles from Stern's Motel. My father consulted the proprietor, Mr. Stern, who seemed to embody his name, with glasses, thinning gray hair, square shoulders, bolo tie. The beach was safe, he said, if we avoided the seedy section to the left of the pier. Off we went, Sally and I, in our jean jackets and dungarees, walking along five-lane Washington Boulevard, the sky blushing pink, streaked with brown smog, the air stale and hot and smelling of exhaust from muscle cars rasping into the night.

At the end of the boulevard we saw the long cement pier and immediately walked to the left, the seedy section, where a few people circled a conga player, his beat dampered by the crash of waves. At the end of the pier, past dried fish guts, a lone fisherman casting, in the public bathroom Sally and I smoked angel dust that we'd smuggled in our luggage, along with shorts and bathing suits. We hitchhiked back to the motel and locked ourselves in the bathroom and smoked more dust, making faces in the mirror, everything distorted. After a while my father knocked on the door and nervously asked, "Have you been taking grass?" We denied it, and he let it go, probably thinking that like many kids, we were experimenting with marijuana.

In the photo of my sisters and me standing in front of the CITY OF STANTON sign, a town north of Anaheim, I look tough in my dungaree jacket. Joanne in her striped tank top looks happy. Sally looks stoned. Barbie was just eleven; her arms are twisted in an anxious pose, even though she's smiling. She must have been worried that entire trip that Sally and I would get caught, her stomach churning with fear, and when at Disneyland the teacup ride was closed, Barbie, with my father, walked as slowly as she could back to where we were waiting, knowing I'd be smoking, and there I was, leaning against a wall, Marlboro in hand, blowing smoke rings. My father walked past, barely glancing my way. "That looks dumb," he said. With three words he reduced me, like in that cartoon — *size of a mouse.*

My mother worked the night shift at Pondville Hospital so she could get the younger kids off to school; Patrick, Barbie, and Mikey were twelve, eleven, and five the year I began my descent. And she picked up float shifts at Medfield State and elsewhere for extra money. How could she keep track of us all? In ninth grade Patrick skipped school for a full month before my mother caught him. He'd don his uniform for Bishop Feehan Catholic school — white shirt, forest-green sweater with a shamrock logo; he'd been kicked out of West Junior High. He'd say goodbye to my mother, then stroll out the side door as if to catch the bus at the top of the street, but instead he'd make a U-turn and like a burglar slip into the basement through the bulkhead. He stowed away in the Orange Room, behind the bar that Ed built, reading Conan the Barbarian paperbacks and peeing in a bottle, my mother upstairs cleaning before she left for work, until one day she found a bottle of urine and figured things out.

Like my father, my mother knew nothing about angel dust. She was a teetotaler; how could she conceive of the drugs easily available to us, drugs that hadn't existed in her youth? One night Paula, Alison, Nicky, and I were in the Orange Room smoking angel dust. No one was talking, because you don't, or can't, talk much when you are dusted. There must have been an album slowly, lopsidedly spinning on the turntable, Neil Young or Pink Floyd or Robin Trower or Lou Reed. My mother knocked on the door, so I opened it a crack, and she tried to hand me a plate of nicely arranged slices of pound cake sprinkled with powdered sugar and some napkins. "We're not hungry," I said. Angel dust was not like marijuana or hash. We had no enhancement of the senses, no cravings. Dusted, I was repulsed by food. I weighed eighty-nine pounds then, anorectically thin from smoking angel dust. The smell of angel dust nauseated me, an acrid chemical odor, the sickening stench of charred leaves, like the smell of burning hair.

My mother pushed the plate into the gap. "I'm sure your friends will like it," she said. I opened the door just wide enough to take the plate, which I set on the floor in the middle of the room, but no one touched it. I suspected Paula and Alison were suppressing laughter about my 1950s-style *Ozzie and Harriet* mom who thoughtfully provided snacks to the teens in the rec room. I was embarrassed by my mother's kindness, a measure of how far removed I was from normalcy. I wrapped a piece of cake in a napkin and threw it in the wastebasket to make it look like we'd eaten some. Somehow in my stoned-out spaciness I was concerned about my mother's feelings; she would be hurt if we didn't eat her cake. At least I was thinking of someone else's feelings and not solely my own.

Where were my parents when I was destroying myself, I've wondered. I see now — like a split screen on television — that my mother, at least, was upstairs baking cake.

On the weekends my mother went to New York to see Ed, we were left alone. Our house earned a reputation as a party house. Nearly every weekend cars cruised down our street, townies and druggies staring out their windows, wondering if the three or four junker cars in our driveway signaled the start of a party. Sometimes the parties grew out of control and our house filled with kids who were not our friends, burnouts like Dan Valerio, Wayne Kosinski, Rod Tyler, and sometimes older and more hardcore townies and dealers, Lighty and his sidekick Woody, news of the party spreading like a contagion, our house jammed, people spilling into the yard, the street. Worse was when kids from Norwood showed up, from the Flats, the tough section. Then we knew we had to call the police on ourselves before a fight broke out, or something worse.

Joanne was fourteen then, and probably with her friends, sleeping at their houses. Patrick, who was twelve, was out with his troublemaking posse of boys, running wild in the projects, the cemetery, the tracks. Barbie, eleven, was upstairs in her bedroom with her door closed, unable to sleep through the noise, anxious because she heard her doorknob turning, someone's hand on it, the crack of light from the hall when the door opened, a drunk stranger stumbling into her bedroom, fumbling around before realizing that she was in there.

On Saturday mornings my father dropped Mikey back home, and we babysat him on Saturday nights. One time Sue had a few friends over while babysitting, and the party grew with Sally's friends and my friends, and then the cars lining the street attracted more people until our house was crowded. One of Sue's drunk friends, a jock with platinum-blond hair, picked up Mikey's small plastic tank of sea monkeys. She couldn't see them, so she didn't believe they existed. Sea monkeys were ubiquitously advertised in comic books or *Mad Magazine*. As a kid, you had to have them, because they grew from nothing, like magic, and in the egregiously deceptive ads, the sea monkeys seemed huge, with tiny cute faces; it was like growing your own family of teensy underwater people.

I had them as a kid and now, years later, Mikey had sent away for sea monkeys, had dissolved in water the dry eggs that looked like powder, had waited weeks, and even though they were just specks, the sea monkeys were visible now through the magnified circles on the clear plastic tank. But Sue's drunk friend couldn't see them, so she held the tank up to her face and stared into the water, then shrieked because now she saw the wriggling creatures, but the tank slipped from her hand and the sea monkeys spilled all over the kitchen floor. I hated her for killing Mikey's sea monkeys.

Some of the kids were drinking vodka and lemonade, so there were half-full plastic cups on tables and counters. Mikey wandered through the rooms, the cutest towheaded boy with big brown eyes — everyone loved him. From across the dining room I spotted him guzzling from a plastic cup that someone had abandoned, a drink that tasted like lemonade, the cup empty before I could stop him. Soon he became wild. The rooms connected around the staircase, and Mikey lapped this circle, zooming around the house like the boy chased by the tigers, like a puppy that raced maniacally when let off its leash. Finally he stopped, and Sue or someone said he should get to bed.

The next morning Mikey walked down the stairs in his underwear and sat on the bottom step and hung his head. "I don't feel good," he said, and cried. He had a hangover — my baby brother, age five. We gave him our cure for hangovers, a big glass of Coca-Cola with ice. I felt sick with shame and guilt. What if he'd found another glass of vodka and lemonade, or a third, and drunk those, too?

Even now I have a sense of worry and guilt about exposing my younger siblings to the parties filled with drunk and stoned and fucked-up kids, kids breaking things, spilling beer and drinks on rugs and furniture, leaving burn marks from cigarettes and joints, throwing up in the bathroom or not making it there in time, kids pairing off to

various rooms to have sex, kids wandering around our house and opening doors, twisting the knob on Barbie's bedroom door.

> November 29, 1975 — Me and Paula and Alison split a gram of dust,
> then went home and I got really cocked. When I woke up, Wayne Ko-
> sinski and Dan Valerio were in the house. Made raviolis at 5:00 a.m.
> and cooked them eggs. It was a totally weird night.

I was fifteen the night Wayne Kosinski and Dan Valerio found their
way into my house while I was passed out. My memory of that night is
olfactory, the smell in the Orange Room: rank body odor, punky socks,
the sweet cloying scent of spilled alcohol seeped into the carpet, the
upholstery saturated with pot stink, the air stagnant in the musty win-
dowless room. Wayne, who was twenty, was one of the biggest angel
dust dealers in town. In three years he'd be charged with intent to distribute PCP, and a couple years after that arrested again for dealing
angel dust. Dan was nineteen. Years after he spent that night in the
Orange Room, after I cooked him ravioli and eggs at dawn, he was convicted of indecent assault and battery on a child under fourteen.

There were summer nights in downtown Walpole that defined being
a teenager in the 1970s, nights when thirty or forty kids milled around
Friendly's parking lot, the sun setting late, heated air rising off baked
asphalt into the cool night, our own weather zone; kids in idling cars,
kids on ten-speed bikes, kids on foot, boys leaning into car windows,
T-shirts slung over bare shoulders, girls in halter tops, backs bare and
breasts loose, hip-hugger pants with three buttons to the bikini line,
midriff shirts exposing skin, boys with red bandannas taming long wild
hair, everyone waiting, someone's tape deck blasting rock music — *and
the radio played that forgotten song* — the thick scent of pot, a Frisbee
scraping the pavement, glancing off a car, the air electrified with a rest-
lessness born from the slow-burning fuses of youth and boredom, a
thrumming energy latent with expectancy, as if we were waiting for
something to happen, a spark to ignite, whispers of a rumble with our

rival town, Norwood, everything vibrating at a higher tension, excitement gathering until inevitably the cops pulled into the parking lot and broke up the crowd.

We'd disperse, only to collect again like minnows schooling, migrating to hidden places as darkness fell to drink and get high, behind Center Pool or United Church or Giantland, a grassy clearing behind Giant's department store, one ancient sprawling oak in the middle of the field. Often we partied at Bird Park, the grand vision of Walpole's most famous industrialist, Charles Sumner Bird, heir to the Bird & Son shingle factory, one of the oldest factories in the country, established as a paper mill in 1795. The park had eighty-nine acres of lawn and woods enclosed by a wrought iron fence, with a stone path winding through like a yellow brick road. There was an open-air amphitheater — I remember as a kid sitting on a stone bench watching actors at dusk — and a creek that fed a swimming pool that was more like an oval pond. In the middle of the pond/pool was a statue of frogs on a lily pad. In summers when we were young my mother packed lunch and my siblings and I spent the afternoon swimming, diving from the frog statue into the greenish water.

At Bird Park my father taught me to play tennis with my new racket signed by Pancho Gonzales, the Mexican American champion, whom I conflated with Speedy Gonzales, the cartoon mouse who was so fast he played tennis against himself, whizzing from one end of the court to the other, faster than the ball. ¡Arriba, arriba! ¡Ándele, ándele! In fall we picked chestnuts there, the seed casings like green apples with spiky shells, which we squished open beneath our sneakers to reveal shiny smooth dark-brown chestnuts, or occasionally the thrill of a pure white one, or a "doubler." The chestnuts had an oily sheen that made them sensual to hold. We collected dozens and at home bored holes in the chestnuts with crochet needles to string enormously heavy and absurdly ugly necklaces.

By the time I was in high school, the pool had been drained because of biohazards and weeds sprouted through cracks in the cement. The amphitheater was burned, its remaining stone walls graffitied. The tennis nets sagged; the spinning octagon ride in the playground tilted on its rusted center pole. Few of the lanterns along the paths worked, so there were dark recesses where we could drink and get high. There was a thrill to roaming around these hidden places, like being invisible in public, lurking in the shadows behind United Church under the noses of the cops, their station a block away, like a map that only we could see with our night vision. We were always looking for a party, though once we found one, the scene was the same, kids drinking and getting high, the faces and places almost interchangeable, even when we crashed a party in another town, like the night in Plainville, which began like any other summer night in Friendly's parking lot.

I was with Nicky when his sister, Andrea, and her boyfriend, Chucky, with a nimbus of fire-colored hair, pulled up to us. They offered us some crossroads amphetamines, little white pills with an X incised into them. I'd taken speed before: black beauties, shiny capsules of biphetamine (similar to Adderall today), one of the amphetamines the military freely gave to soldiers in Vietnam for pep, 225 million tablets distributed in just three years of the war; and crystal meth, a $10 tinfoil packet filled with granular white powder that we snorted, drinking beer to temper the jittery amped-up buzz. When a batch of crystal meth came into Walpole, parties were filled with kids grinding their jaws, a twisting motion like ungulates chewing. Monica Lande, a cool older girl who rarely deigned to speak to me, cornered me one night after she'd snorted crystal meth, her pupils dilated as she talked and talked, her jaw sliding sideways.

Nicky and I swallowed the speed, and then Chucky said he knew of a party in Plainville, so he drove us there. I don't remember much about the party, just kids I didn't know sitting around drinking, smoking pot, getting wasted, rock music so loud you couldn't hear anyone talk. Nicky sat in an easy chair and I curled on his lap, my cheek against the soft skin of his neck, breathing in the clean talcum scent of him. Later we walked upstairs to a bedroom and crawled into a bed. We took off our pants and shirts, and in some stranger's bed beneath some stranger's sheets we kissed and touched, but soon we fell asleep. Probably our drug and alcohol habit prevented me from being a pregnant teenager.

The next morning was bright and sunny and we woke up surprised but happy to find ourselves in that bed in that room, fuzzy with sleep and cotton-mouthed. Downstairs I could see now that the house was contemporary, with huge picture windows overlooking a yard with jack pines, a sand driveway — a landscape I'd seen only on Cape Cod. I drank a glass of water, but there was nothing to eat. I couldn't eat anyway. Sometimes with Nicky I felt so in love that I became nauseated, an overwhelming physical sense of desire that resembled anxiety, like an overdose of hormones and emotion. One night earlier that summer we swam in the pool after midnight, everyone else out or asleep, moonlight glinting off the water as we pulled off our bathing suits, the water cool and almost oily as we drifted toward each other and I wrapped my legs around his waist, our skin slick and warm as we pressed together, floating, as weightless as our consciences.

Nicky and I were alone in the house. Even Andrea and Chucky were gone, but surely they'd be back to pick us up. The living room was cluttered with empty beer cans and bottles from the night before, ashtrays overflowing with stubbed-out cigarettes. I hunted in ashtrays for snipes, cigarette butts with a bit of tobacco left. Nicky explored the house, opened cabinets. We had no idea who owned this house or where we were exactly, just: Plainville.

I followed him down to the basement, to a workbench cluttered with materials to make rifle cartridges, like the spent shells I collected in the woods behind my house, red and green plastic tubes the size of a Bic lighter, with brass caps. There was a tin of gunpowder and — as if the owner had been called away in the middle of his work — a pile of black powder on the table. As kids we played with gunpowder in the form of caps, a half-inch-wide ticker tape of red paper pocked with freckle-sized blisters of black powder. We'd lay the strip of paper flat on the sidewalk and smash a rock onto each blister until it exploded. Over and over we smashed, thrilled by the tiny crack of the explosion, the puff of smoke, the strangely appealing whiff of sulfur like rotten eggs, the small white star-shaped scar on the asphalt from each bang.

We'd buy a box of five rolls, 250 caps, and we'd bang and bang. There were guns into which you could load the caps, but I didn't have one. Guns were for boys, not girls. We graduated from caps to cherry bombs, bottle rockets, firecrackers, M-80s.

Nicky poured some of the gunpowder into a baggie and stuffed it in his shirt pocket.

"What are you going to do with that?" I asked.

"Make a bomb," he said. "Put it in the school." He laughed and I laughed with him, and we walked upstairs.

I didn't think Nicky would make a bomb; I doubted that he knew how. Occasionally kids called bomb scares into school to try to get classes canceled. How did kids even know to call in bomb scares? The 1970s was a bomby decade. In 1972 alone there were over 1,900 bombings by radical activist groups—the Weather Underground, the Black Liberation Army, the Chicano Liberation Front, Americans for Justice, the FALN, a Puerto Rican nationalist group, the Jewish Defense League, anti-Castro Cubans, Croatian separatists. During an eighteen-month period in 1971 and 1972, the FBI reported more than 2,500 bombings on U.S. soil, nearly five per day. At the height of the domestic terrorism in the country in the 1970s, there were a thousand bombings a year.

I didn't learn about the radical underground movement in "America in the Twentieth Century," at least not in the classes I managed to attend. But the blasts rippled into our psyches. I didn't learn of the Weather Underground or the Black Panthers, but I absorbed that angry energy, glimpsed it on television and in magazines, that unforgettable powerful image of Patty Hearst wielding a machine gun, a *girl* who'd joined an underground army, the Symbionese Liberation Army, which I thought was trying to liberate some desperate people in a foreign country—the Symbionese, like Lebanese or Sudanese, the underground army's headquarters actually underground, I imagined, like the Batcave, which was why they were hard to catch. That's what I gleaned from headlines.

Two months before Nicky took the gunpowder, in April 1976, a bomb exploded in the Suffolk County Courthouse in Boston, injuring twenty-two people. The bomb was planted by the United Freedom Front, which aimed to obliterate the parole office and its records tracking thousands of ex-cons. The United Freedom Front was birthed from the prison rights movement, spearheaded by Raymond Levasseur from Maine, a Vietnam veteran who'd been sent to a maximum-security prison on a minor marijuana charge, and Tom Manning, also a Vietnam vet, from a working-class Boston Irish family. After returning from Vietnam, Manning robbed a liquor store and was sent to Walpole Prison for five years, where he was stabbed by another inmate and nearly died. "I spent my last fourteen months in Walpole's 10 Block, where I read Che, and where all the prisoners — black and brown and white — were united out of necessity," he wrote. In a letter about the bombing addressed to the *Real Paper,* Boston's alternative weekly, the United Freedom Front wrote, "This is but the sound before the fury of those who are oppressed." They demanded prison reforms, including an "immediate end" to "the 23½ hour a day lockup" at Walpole, enforced on some inmates for nearly two years.


Bombs exploded inside Walpole Prison, too. In 1972, Stanley Bond, a member of the Weather Underground incarcerated in Walpole for robbing a bank, was killed when a bomb he was making in the prison foundry detonated. And in April 1975 another homemade bomb exploded in the prison, also killing the inmate who'd been holding it.

Bomb scares weren't abstract or in cities or behind cement walls. In the town of Walpole, forty-two bomb scares were reported in the local paper in 1976, almost one weekly. (There were only two in 1977, and none for the rest of the decade.) To ring in 1976, on New Year's Eve three bomb scares were called in to Walpole police, one for the police station itself, two for restaurants downtown. Two weeks later, "an unidentified caller" told police a bomb would go off in Papa Gino's pizzeria, where I ate twenty-five-cent slices. In February five bomb scares were called in to the police within ten minutes, one for Friendly's restaurant, the others threatening local factories.

In the following weeks there were bomb threats to On Luck Chinese restaurant, a mile from my house, and Walpole Bottling Company, where as a kid I went with my father to fill a twenty-four-quart case of soda, picking flavors from the warehouse floor, cream soda and sarsaparilla and lemon-lime. There were so many hoax bomb threats that the police reaction was blasé. There were "no major crises" over the bicentennial weekend in July 1976, according to the police log, just "summer mischief" — a streaker in Bird Park, a sighting of a large red UFO, and two bomb scares, one at Almy's department store, a favorite shoplifting haunt, and one man warning that a bomb would go off in the town hall. Streakers, UFOs, and bomb threats — strange days.

The bomb scares in Walpole over the bicentennial were hoaxes, but real bombs exploded that weekend elsewhere in Massachusetts, set by the United Freedom Front, again demanding prison reforms. They detonated bombs at a National Guard armory south of Boston, the Essex County Superior Courthouse north of Boston, under an Eastern Airlines prop jet at Logan Airport, and outside a First National Bank on the North Shore. The plane and buildings were damaged, but nobody was hurt.

The United Freedom Front, "the last revolutionaries," continued bombing throughout the decade and into the 1980s, mostly in Massachusetts and New York, until they were caught in April 1985. Then the men who'd been radicalized in prison, who exploded bombs to demand prisoners' rights, wound up back in prison.

I didn't worry about Nicky using the gunpowder for a bomb. He wasn't violent or aggressive, and I never heard him express political views. He was looking for thrills, rebellious but not angry. He was at heart a *nice boy*. (He told me once that I should be kinder to my mother; he was close to his.) The gunpowder in someone else's possession might have been dangerous, but the appeal for Nicky, I imagine, was to hold in his hands the potential for explosive power, like a kid smashing caps on the sidewalk.

• • •

Chucky and Andrea returned to the house in Plainville; they'd just gone to the store for smokes. "Let's get out of here," Nicky said. He didn't want to get caught with the gunpowder. All day Saturday we drove around getting high, Nicky and I leaning into each other in the backseat. Maybe we took a second hit of speed; I don't remember stopping for food, or eating. Maybe that's why we couldn't seem to stop driving and going nowhere, driving and circling and getting high and listening to the car stereo. By the time it grew dark Saturday night, we'd been out for more than twenty-four hours and we were running out of gas and money and drugs. My mother was at work, so we planned to drive to my house, where I had a few dollars and some pot stashed, but first we needed gas.

Chucky had siphoning equipment in his trunk. He drove down a dark rural road until he saw an unlit house with a car in the driveway close to the road. He gave Nicky instructions, and Nicky disappeared down the street carrying a red jerry can and tubing. He returned five minutes later, gagging and spitting but with a gallon of gasoline, minus the mouthful he'd swallowed, and we drove away in Chucky's Pinto, the Ford model that tended to ignite when rear-ended, Nicky and me in the backseat, he with gasoline in his mouth, gunpowder in his pocket.


As we cruised through the center of town, Paula flagged us down. She poked her face in the window. "Where the fuck have you been?" she said. "Everybody is looking for you. You're screwed." She nodded at Nicky and me in the backseat. "There's an APB out for this car." All points bulletin. How did she know this? Maybe Good Goin' Gus, the friendly cop, told her. We'd been reported missing, as runaways, by Nicky's mother when he and Andrea hadn't come home Friday night. With the stolen gas, we drove to my house and I ran in for the $3 and the pot. On my way out, as I bolted down the stairs, I bumped into Sue in the kitchen. "You'd better stay in," she said. "Mom knows you didn't come home last night. She knows you weren't at Alison's." My mother would be home from work soon. I walked past Sue. "Don't be an asshole," she said. I walked out the door.

In the car, I told everyone we were in trouble. Nicky vowed never to go home, and I was with him. That option seemed easier — not having a place to live, no money or food or clothes — than facing parents. We'd just do this forever, drive around getting stoned, driving and driving and never arriving. *It's all the same fucking day, man,* as Janis Joplin said. She was my hero, the voice of pure anger, desire, defiance. Like Janis, I was hardened, tough — you can't hurt me because I have *nothing left to lose.* Bent on self-destruction, I didn't care about my future because *tomorrow never comes.*

We bought gas and smoked the pot and soon again we had an empty tank and an empty pipe. Chucky said he was taking us home. He was Sue's age, seventeen. He stood to get in more trouble than Nicky or me; maybe he realized this, or maybe he realized that there was nowhere to go, no choices left. When I walked into my house after midnight, my mother was waiting for me, still wearing her nurse's uniform, pacing in the kitchen. She was angrier than I'd ever seen her. I tried to brush by her, but she grabbed me by the collar and pushed me against the dining room wall. I don't remember what she said. What could she say? "Let go of me," I said. She gripped my shirt, her face an inch from mine, though she was shorter by three inches. "Let go or I'll walk back out and never come home."

"If you do that, you will never step foot in this house again." We both threatened the thing we didn't want to happen.

"You wouldn't have missed me if Nicky's mother didn't call," I said.

This bothered my mother, this accusation of lax parenting. I just wanted to go up to my bedroom and sleep. She probably just wanted that, too. But it was like the Bay of Pigs. She had to assert her power and authority over me, and I couldn't let her, so instead we threatened annihilation. Finally she released me and I walked upstairs.

The next day she told me I was grounded. I laughed. She'd lost me; she must have known already. I'd lost myself. I just didn't know it yet.

·   ·   ·

Another Saturday night, another party at my house, though by 4 a.m. most of the people had crashed or left. Alison and Nicky and I were still awake, so we decided to take Sue's car for a spin. None of us had our licenses, but there wouldn't be many cars on the roads. Nicky drove, steering us along back roads where cops were less likely to lurk, and we found ourselves at the golf course where he caddied. The entrance was a narrow road between hedges, like driving through a tunnel, but then the hedgerow opened to moonlit swells of lawn, a parking lot, and in the distance the darkened clubhouse. The place was silent, inert. Nicky knew the landscape from hauling golf bags across eighteen holes, knew that when he suddenly swerved left and gunned the engine that we'd career over manicured lawn, knew that when he yanked the steering wheel hard to the right that he was routing the fairway, the tires spinning and ripping up turf, gouging the delicate putting green, revenge against the wealthy men whose heavy clubs he lugged for hours on hot days, like a servant, a footman, but who failed to tip him.

Nicky spun the car, laughing and whooping as he struggled free from ruts, clumps of turf spraying off the back tires. Alison and I were tossed across the vinyl seats as if we'd taken a wave across the bow, Alison in front, me sliding around the back, which might have symbolized my ambivalence about this plan — I was a participant in adventures, a sidekick, not the driver but along for the ride. At least I remember sitting in back, but that could be a trick of memory to assuage my guilt. After all, I'd found the keys, I'd given them to Nicky, I was there. Nicky spun another doughnut, then sped across the lawn. I saw a glow on the horizon like waking from a dream. "Nicky, we have to get out of here. Now!" Employees and caddies and early-bird golfers would arrive soon — any minute, it seemed. Nicky aimed the car out the narrow driveway and we sped away, exhilarated, horrified.

A couple miles down the road, the engine quit. Nicky coasted to the shoulder and tried to start the car, but the key yielded an unpromising *click click click*. We realized how close we'd come to the car stalling on the golf course amid our wreckage, how narrow our escape, and Nicky still on probation from the B&E. It must have been 5 a.m. We knocked on the front door of a house and a woman let us inside to call a tow truck. We thanked her, and outside Nicky tried one more time to start the car, and miraculously the engine turned over, so we drove off quickly, passing the wrecker in the opposite lane a half mile down the road. We pulled into my driveway, chunks of sod hanging from the front bumper, a faint burnt odor, the overheated engine ticking.

Destructive urges in little kids are tests of power — pulling wings off insects or stepping on ants, sweeping a mighty arm across a game board, scattering pieces in a fit of frustration. As teenagers with no direction or oversight, no toys, our destructive urges arose out of malaise, unfettered time, boredom, and no connection to anything — not to school or family, not to the town or clubs or sports, no purpose or aim, just pent-up energy, the need to release that energy, to smash and burn.

> Nov. 10, 1975 — Nobody can party up the cemetery anymore. Someone broke the tombstones.

In 1975, Walpole police were "besieged" with vandalism and a "major rampage of window smashing." Over four days, more than sixty windows were broken at schools and stores. One quiet night as a group of eight or so kids walked through the deserted center of town, Derek Lowery spontaneously karate-kicked the plate-glass display window of Betro Pharmacy and it shattered, raining glass on the sidewalk, surprising even Derek. We dispersed like startled birds, fleeing to the tracks behind the block of stores, seeking each other afterward, dark figures stepping over railroad ties — *Is that you?*

Alarmed by the rise in crime in town, which paralleled a national trend, Walpole hired six more police officers, bought two new cruisers, and deployed an armed security guard with a patrol dog to schools, a favorite target of vandals. The town imposed a curfew on the commons to prevent "rowdyism" — anyone seen in the town center past 9 p.m. was fined. The curfew lasted all summer, though it only drove kids into the shadows, to congregate in woods and abandoned lots, gravel pits.

If 1975 was the year of smashing windows in Walpole, 1976 was the year of setting fires. In the first three months there were 500 fire alarms, more than typical for an entire year, with "vandals responsible for most of them." In just one week in April there were 122 fire alarms and 90 fires "of suspicious origin."

The fire my friends set one day was suspicious, but more accident than arson. Six of us sat in the woods adjacent to the train tracks on a Sunday afternoon, just beyond the town center, three boys and three girls. As we passed bowls of pot, the boys idly tossed lit matches into the dry pine needles. They made a game of it, allowing the two-inch-high ring of flames to eat away at the duff, the patch of scorched earth spreading before they raced to extinguish the fire, until one time they waited too long and the fire caught a breeze and surrounded us, and then we all stomped madly, half panicked, half thrilled, an acrid smell of melting rubber from my work boots, the hem of my jeans singed, the smoke thickening and rising above the trees, prompting someone to call the fire department, the sirens growing closer, louder, as we ran down the tracks, the soles of my shoes smoldering.

I never tossed a lit match into the dry tinder of a field or woods or weighed the heft of a rock before pitching it at a window, even though I was present with boys who did. Like most teenage girls, my destructive tendencies were aimed inward; we were capable of destroying ourselves. *You can't hurt me, world, because I will hurt myself first and best.* Alison scraped lines on the inside of her forearm, in the soft pale flesh that never tanned. In the bathroom or in her bedroom she toyed with razor blades, etching her flesh, summoning pearls of blood, blood turned from blue to red, freed to flow outside instead of in the crazy endless loop in her body. Maybe the cutting was to scarify, to break open a surface, like a medieval bleeding to release bad spirits, a letting to relieve pressure. I tried cutting once to see what it felt like, to understand why she did this, scraping the point of a safety pin along my wrist. I wasn't serious — I'd used a *safety* pin. The act felt melodramatic and conspicuous.

I wanted to *not* feel, so I smoked angel dust. Alison's and my behaviors were two sides of the same coin. My self-destruction was less deliberate: not caring about myself, my health, my safety, leaving my fate to others, drinking and taking drugs and riding in cars with people who were so fucked up they could barely walk, let alone drive, getting behind the wheel of a car myself in that condition, tempting fate. Alison was interested in thanatology, she said, the study of death. She wanted to be a thanatologist, if one could be such a thing. I wasn't sure she knew what a thanatologist did, but it was a defiant word: *I am not afraid of death; I will study death; I will master it*. It was like when I announced to my friends in junior high that I was an atheist — like saying *fuck you* to the most powerful entity imaginable then. *I don't believe in you, Mr. God. I don't need you. Put that in your fucking pipe and smoke it*.

Maybe Alison and I and any of us didn't know what to do with our energy, our power, the exuberant adolescent rush of hormones, thrill the most alluring drug. Or what to do with our engulfing sadness, our confusion, our anger. I was filling up with rage then, though I barely knew it. Maybe coming close to death was a way to excite life, to live on the edge, to literalize that edge, cut with an edge, to see your own blood and ask, *Am I alive?*

One spring night Alison and I and a few other kids sat behind United Church smoking pot, drinking. The church lawn was encircled by evergreens, and though the police station was just beyond the trees, cops never checked the field. That night after the sun set, Alison whispered that she had to go to the bathroom. She walked toward the middle of the field until blackness swallowed her. She was gone a long while, and we began to wonder. "Maybe she went home," someone said. Then I heard her call my name, weakly: "Maureen." Everyone laughed. They thought she was summoning me to help her pee. I stumbled through the darkness and nearly tripped over her lying on the grass.

"What's wrong?" I asked, crouching beside her.

She clutched her stomach, rolled to her side. "I'm think I'm dying," she said.

My heart pounded. "What did you take?" She'd tried this before, which was why I knew to ask. Pills, she said, stolen from Betro Pharmacy.

"How many?" I must have been trying to decide if she was just going to throw up, or maybe something worse, but how would I know?

"The whole bottle."

*Jesus Christ.* I shouted for the others, and when they reached us, someone said, "We better call the police."

"No," Alison moaned. "Please don't call the police. I can't let my parents know."

We promised Alison we wouldn't call the police. There were no cell phones; the 911 system was years away. Two of us ran the couple of blocks to a phone booth on Main Street to call Project Face; their flyers offering counseling for teenagers were plastered around town. Someone on the other end of the line told us to call an ambulance. *But we can't let the police know, or Alison will be in trouble.* The person said to get Alison to a hospital right away.

When I think of the time we wasted — running to the phone booth, finding coins to make a call, dialing the excruciatingly slow rotary phone, the line ringing, then the click of someone picking up, explaining to the counselor, time passing in the watery slow motion of a nightmare while the pills, whatever they were, dissolved in Alison's bloodstream — I feel sick with the possibility that we might have cost Alison her life; it scares me still.

We flagged down Spider D'Amico cruising along in his mother's station wagon, and he drove behind United Church and we laid Alison in the backseat, moaning and clutching her stomach. We jumped in and Spider sped down Route 1A, all of us urging him to hurry, to run the red lights, because Alison was eerily silent now. She didn't respond when we lightly slapped her cheeks. Her eyes were closed, her eyelashes dark and thick against her smooth skin, jaundiced under the car's dome light, her mouth slightly open, her lips like she'd eaten a cherry Popsicle. It was strange to see Alison so slack, contrary to her usual vitality, her raucous laugh, that determined way she walked, with quick short steps. In ninth grade she'd been voted most popular, a charmer whom boys dreamed of, girls imitated, envied, but now she was slumped in Spider's car, listless.

Spider swerved into the emergency entrance of Norwood Hospital and two men in scrubs lifted Alison and rolled her down the hall on a gurney. In the waiting area, a doctor approached us and said they needed to pump Alison's stomach but they couldn't without permission from her parents, so we must tell them Alison's name. We'd promised Alison, so nobody spoke. The doctor looked at one of the boys. "Come with me," he said. When the boy returned a few minutes later, we asked him if he'd told, and he nodded. "She's screwed now," someone said. The doctors said that Alison would die if her stomach wasn't pumped, but they couldn't proceed without her name. We quieted then, all of us falling for the doctor's ruse, relieved that Alison would be okay.

Alison was admitted to Westwood Lodge, a private psychiatric hospital five miles from my house. Westwood Lodge was where the poet Anne Sexton had stayed after her suicide attempts. In October 1963, Sexton wrote to her husband from Rome: "Darling, I need therapy . . . Need even (god forbid) Westwood." In Westwood Lodge, that "sealed hotel," Sexton wrote poems for her first book, *To Bedlam and Part Way Back*. There had been many unfamous residents of Westwood Lodge, like my aunt, who had what was called in the 1970s a nervous breakdown, as had the mother of a friend of Patrick's, and kids from Walpole who'd smoked too much dust.

Every once in a while you'd hear of someone "freaking out" and then that person disappeared for a week or two, or a month, often to Westwood Lodge: Michelle Laski, Tammy Hurley, Gary Gravino. One day I stopped to talk to Gary in the parking lot after he returned to school from Westwood. He smiled, but his speech was thick, his eyes glassy. I noticed cookie crumbs around his mouth, caught in the fine hairs above his lip, but he seemed unaware of the crumbs. His vacant eyes, the sad crumbs on his face as if he were a child, broke my heart.

Walter Slater freaked out and disappeared for a while to Westwood.

He'd crashed in the Orange Room one night after a party, so high and drunk he fell asleep with his arms around his Frye boots, those clunky, round-toed quasi–cowboy boots that were part of a leather fad in the 1970s — leather coats, leather boots, leather vests, leather satchels for your pot that cinched closed and hung from your leather belt. "I love my boots," Walter mumbled that night. Walter Slater was a dust-head, a small-time dealer, but he got clean at Westwood and was working as a drug counselor. Later, when I saw him smoking angel dust again, I said, "I thought you quit." He did, he told me, but he felt like a hypocrite. Every minute he was telling some kid not to take drugs, he felt an intense urge to get high. So he did.

Paula and I visited Alison in Westwood Lodge, driving slowly past the manicured grounds to a manor-like house built in the 1920s, with ivy creeping up stucco walls. There was a newer, clinical-looking building as well, but the facility as a whole was small — ninety beds. I don't recall having to sign in, but surely there must have been some way to register visitors, or a house phone on which we called for Alison. I remember the old-fashioned parlor where we sat, large, with ornate trim and high ceilings, arrangements of wing chairs and settees, end tables with lamps that shed amber light. Floor-to-ceiling windows overlooked a lush lawn and beyond that a thicket of trees. The kitchen had a butcher-block island, stainless-steel counters, an industrial-sized refrigerator. Alison stood in front of the refrigerator as if this were her home, grabbed a drink, took an apple from a bowl. "Have one," she said, sharing her bounty, but I didn't want to eat an apple from that place.

In spite of the plush chairs and couches in the sitting room, the only other person there was a tall, thin, stupefied boy with messy blond hair and glasses, who followed Alison around, who loved her, it seemed, as did so many boys. Alison spoke directly to him — "Ted, go watch television" — sometimes rudely — "Ted, go play with yourself." At least she spoke to him, and that seemed to please him. I sensed that Alison enjoyed — at least a little — her stature in that milieu, "queen of this summer hotel," as Sexton had called herself at Westwood. There was a certain exotic glamour to being sent away to an institution, to being incorrigible.

*Incorrigible* means "beyond correction" — broken, unfixable, a hopeless case, recalcitrant, intractable, obstreperous, truculent, insubordinate, defiant, rebellious, willful, wayward, difficult, troubled, delinquent, deviant, miscreant, stubborn. "Stubborn" was a diagnosis in Massachusetts until the early 1970s, and boys and girls as young as seven were sent to reform schools for being stubborn children. Nonpejorative words to describe such kids might be *spirited, unconventional, bold, assertive, extroverted, adventurous, risk-taking, daring, verbal, sensitive, artistic.* Or words that describe moods or stages of adolescence, temporary — *sad, angry, insecure, confused, depressed, fearful, bored, frustrated, anxious, alienated, lonely.*

I didn't see any orderlies or nurses at Westwood that day, as if there were no supervision, as if the place truly were a lodge and Alison there on vacation. Because of the lax security, we took Alison out for a drive, though we were not supposed to leave the grounds; there was a sense of breaking her out. We drove aimlessly along the country roads surrounding the lodge, smoking pot and probably angel dust — I recall Alison wanting to get dusted. We had no qualms about getting high with Alison, because she seemed normal, not perturbed. We never cautioned each other about the dangers of smoking dust, about excess or self-destruction, though we were aware of some risks; we'd heard the lore of kids drowning while dusted, that in water you became disoriented, lost the ability to distinguish up from down. But we didn't apply the risks to ourselves. We had a sense of time unending.

Nobody talked about that night behind United Church, the pills Alison took.

Alison was admitted to Westwood Lodge again, but the next time she was housed in the newer building in a locked room. We had to ask a nurse for access. In that room with a single bed, a Judas window in the door, Alison was lethargic. She spoke slowly, deliberately, her actions halting. She'd been dosed heavily with Thorazine, an antipsychotic, and Stelazine, an antianxiety medication. Outwardly the drugs had a similar stupefying effect to angel dust's, but it was disturbing to see Alison in that condition, because the drugs were imposed on her. On that second visit, it was clear that Alison wanted to get out of there; that time, institutionalization was not like a stay in a dormitory. But there was no escape from this unit. We could do nothing to get Alison out, and soon a nurse told us we had to leave.

Tranquilized, Alison was fundamentally changed, her energy drained, her speech sluggish. What was the logic in giving incapacitating drugs to cure a girl of her drug problem? Of depression? Alison moved from self-medicating with street drugs to sanctioned "therapeutic" drugs, but either way she was stoned out of her mind. When she returned to school, with scars like vapor trails on her arms, people were shy with her, acted awkward around her, as if she were a ghost or a movie star.

# 7
## *Work-Study*

A YEAR OR SO AFTER NICKY ASKED ME TO GO OUT WITH HIM "FOR as long as it lasts," we exchanged Christmas gifts on our second Christmas together, sitting in Sue's car parked in the driveway, a quiet place where we could be alone. Nicky gave me a ring he'd bought at Zales jewelry store at the mall, a diamond chip set in a gold band, a heart cut into the bezel, and a plush sweater, soft as rabbit fur, which I never wore because it was *too nice;* I didn't want to ruin it. After I slipped the ring on my finger, I started to cry. "What's the matter?" Nicky asked. I was embarrassed by my paltry gift for him, all I could afford without a job. He laughed, then kissed me. "Goofball." He opened his gift from me, a braided leather belt, which he proclaimed to love and wore faithfully every day thereafter.

Later, entwined on the couch in the Orange Room, we decided that the ring should mark our *pre*-engagement, like a category we invented, the ring symbolizing more than friendship, less than a vow. We set a date two years after we graduated from high school — we were practical dreamers — which I wrote on the wall in my bedroom: August 1980, four years away. We were confident that we'd love each other in four years as we did in that moment, that nothing would change, not even ourselves, that in four years we would still be madly in love and want to spend the rest of our lives together. The tentativeness of our plan — four years away, a *pre*-engagement — might have told us something.

• • •

One day in Fields Hosiery at the mall, I stepped up to the counter to pay for a pair of socks. The saleswoman rang them up and then said, with a look of disgust, "What else do you have?"

"Nothing," I said.

"I saw you put them in your jacket," she said.

In that moment I had to decide whether I should continue denying the theft, which might inspire her to call security, or own up to it. I pulled three pairs of socks from inside my coat, threw them on the counter, and walked out of the store, indignant (as if she had infringed on *me*), a cover for my embarrassment. Still, I'd gotten away with one pair of green-striped knee socks stuck up my sleeve. I wore them to school, though each time I tugged them on I felt a flash of shame, remembering the cashier's contempt.

Maybe because there weren't consequences that first time I was caught, I didn't stop shoplifting until Paula and I were nabbed stealing a $2 box of clay from a discount store. We didn't want to spend $2 on something we had to buy for school. The store detective marched us into a small office and threatened to call the police. Should I be sent up for a $2 box of clay, petit larceny? He called our parents instead. My mother was upset, at me and at herself, it seemed. How could she teach me a moral lesson about stealing? She and Ed had been caught earlier that year in Roxbury, a high-crime neighborhood of Boston, where they went expressly to shoplift. They must have thought Roxbury would be a good place to steal, a place where crime was normal, but if crime was higher there, the store owners were aware, equipped. My mother and Ed were thrown out of the store, lucky to avoid arrest.

My mother, Ed, Paula, me — we were four of four million shoplifters caught in 1975. After my mother graduated from the nursing program that year and began to work full-time, she stopped shoplifting. I stopped, too, in December of my junior year. After passing my driver's license exam, I set out to find a job. I applied at Betro Pharmacy downtown, but the old man at the counter said he wasn't hiring. I asked if I could fill out an application, just in case. "We only hire pretty girls," he said. I walked out feeling awful, but then I thought about his employees — what was he talking about? *Fuck him*. Maybe he'd seen me hanging around Mimi's Variety or Friendly's, which bespoke a reputation: troublemaker, drug user, delinquent. He would have been right.

I drove to the mall and applied at nearly every store: Foxmoor Casuals, India Imports, Bradlees department store — my former shoplifting targets. I walked into every pizza joint and restaurant along Route 1, but nobody was hiring, at least not hiring me. On a whim, I pulled into a gas station where I'd once seen a girl pumping gas and asked the proprietor, Henry Delaney, if he was hiring. Henry was a short man with light-brown eyes that fixed on my chest as he handed me an application. He wasn't hiring, he said, but just in case, he had my number.

A week later Henry called to offer me a job. No interview, no questions; I was hired over the phone to work 3 p.m. until 10 p.m. Monday through Friday, thirty-five hours a week, with some Saturdays and the occasional Sunday, for $2.50 an hour, twenty-five cents over the minimum wage. I hit the jackpot, I thought.

The gas station was situated at a busy intersection, with two service islands on either side. Junked and crushed cars were stacked behind the building, and along the front was a triangle of grass where on weekends a man sold velvet art paintings — Elvis, Jesus, dogs smoking cigars — none of which I ever witnessed anyone buying. The station had a glassed-in lobby with a dirty tile floor, a cigarette machine, a row of orange vinyl seats welded together like bus station seats. Behind the service window was Henry's tiny office, just big enough for a desk and a swivel chair, and then a three-bay garage.

I wore jeans and a hooded sweatshirt beneath a puffy goose-down jacket to my new job that winter, pumping gas, checking oil, measuring tire pressure with my handy gauge, squeegeeing windows, checking brake and transmission fluid. Often men pulled up to the pumps and said, "Honey, would you get one of the boys to check my oil?" When I wore a hat, customers said, "Son, can you clean my windows?" or "Fill it up, son." I didn't bother to correct them.

The old man from the pharmacy, who hired only pretty girls, was a regular. The first time he pulled in, he looked at me with a haughty grin. "I see you found a job," he said, as if I'd gotten the job I deserved, not in the bright clean pharmacy with all the pretty girls but outside pumping gas, hands stained with oil, smelling of petroleum, dirty. Pump jockey. Grease monkey, people called me, but I was proud. Girls didn't do that kind of job. After that the old man seemed to take a perverse delight in tormenting me, writing as slowly as possible on the credit card slip, especially on rainy days as I got drenched waiting for him. I grew bold and retaliatory. When I saw his car pull up, I'd finish my cigarette, luxuriously blowing smoke rings, staring out the lobby window at him staring at me. I'd stroll leisurely to his car, where he'd act annoyed at my poor customer service even as he relished my servitude.

In the photo Sue shot of me at work, I see the beginning of a vertical crease between my eyes that already marked my expression as serious, tough — a protective veneer. My hair is parted in the middle, as was the style (or lack of it) in the 1970s, and pulled into a ponytail to straighten the curls. There's a bleached orange patch in my hair from dousing my head with Sun In, a blotch of burnt copper instead of highlights. I look thin, my face drawn. I'd been smoking angel dust for more than a year; I weighed just ninety pounds.

In warm weather I rode my bike three miles to work, then home at 10 p.m., the streets empty and peaceful. I'd zoom through the intersection downtown and coast down the middle of wide, smooth Robbins Road with no hands. I was never late or absent from work as I was at school; I'd been warned that if I missed one more day of school in my junior year, I'd be held back. The threat was effective, not because I worried about my academic performance but because I'd be mortified to be in the same class as my younger sister Joanne.

What's surprising is not the number of days of school I missed but that I managed to attend the rest of the days, though often I was present in body only. Sometimes I'd drive to school and in the parking lot, where kids hung out between classes, I'd crawl into the backseat of my car to sleep. Later I'd forge my mother's signature on a note. One day I woke up to faces smushed against my windows, kids laughing and staring at me curled up in the backseat as if I were one of Mikey's sea monkeys in that plastic tank.

In the small hierarchy of the gas station, below Henry were the mechanics, who drove souped-up cars with dual exhausts and raised rear suspension. They wore leather bomber jackets or jean jackets stitched with emblems — STP oil, Champion spark plugs. They never came clean, no matter how hard they scrubbed their hands with the slimy goo at the basin. Their fingernails retained grease, and the webbing between their thumbs and index fingers was grayish; I know because my hands never came clean either.

My mother bought Sally and me a 1968 two-door Opel Kadett, which we called the Egg for its oval shape and off-white exterior. The car was so small and lightweight that while I was in class one afternoon, some boys lifted the car and turned it sideways in its parking space. One night at work the tow-truck driver installed a cassette deck in my car and shorted the electrical system. The next morning the mechanics stood around the Egg stuck in the bay with its hood open. "I've seen bigger engines on a sewing machine," one said. "What is it, a two-cylinder?" They chuckled at my weird little foreign car before fixing it for free.

Below the mechanics were the tow-truck drivers. Fast Freddie, the daytime driver, was tall and ovoid, shaped like one of those blow-up clowns weighted at the bottom that bobbed upright after you punched it. Fast Freddie moved at one pace, like a three-toed sloth. His very slowness made him cool; nothing ruffled him. Tom Parisi, who drove the wrecker at night, was the opposite of Fast Freddie. Tom was tall and ropy, with nervous energy. He had a slight — and I thought sexy — lisp when he talked. He had deep-set brown eyes and a curved honk of a nose. His face looked chiseled from rock, with sharp angles and hollows. In the lobby while we waited for someone out there on the humming highways to break down, run out of gas, or crash, waited for someone's luck to run out, Tom detailed moneymaking schemes, his eyes wide with the thrill of his plots. One involved changing his license plate numbers with black tape, filling his tank at an outlying self-service gas station, and then speeding off without paying.

My boss, Henry, was a runty boyish-looking man who could be friendly, even fatherly, or, more often, lewd. Once he helped me analyze "Howl" for my English class, and another time he explained paganism. "See, I'm pretty smart," he said. "Surprised you, didn't I?" He did surprise me the first time he grabbed my ass or reached for my breast, then chuckled like he'd won a game of naughty tag, though only he was playing. I learned to skirt him, to slink by, never to place myself between Henry and a wall.

One Sunday afternoon when the tow-truck driver was out on a call, Henry called me into the repair bay. "Mo," he said.

I appeared in the threshold of the garage, about twenty feet from him.

"What?" I thought he was finally going to show me how to use the tire-mounting machine so I could fix flats.

"Come closer," he said, a weird grin on his face. When I was six feet away I noticed his penis pushed over the top of his gray Dickie work pants. "That's just the tip of the iceberg," he said, smiling. Henry had a beautiful smile, straight teeth that glowed white against his tanned skin. I turned and walked outside to the pumps, where I kept busy sweeping up trash, wiping down the glass over the meters, until the wrecker driver returned and Henry left. For a while after this, Henry acted odd around me. He wasn't apologetic but aloof, maybe even a little hurt, or at least annoyed. He had not elicited the response he wanted, which was what? Appreciation? Laughter? Running away screaming like a little girl? After that, Henry tempered his lascivious impulses, at least around me.

Walpole cops pulled in daily, as did state troopers from the nearby barracks, wearing immaculate creased jodhpurs and black boots. Henry fixed their cars, maybe for free, some quid pro quo for which Henry, and by extension his employees, received immunity, a veil of protection. One night in the alley behind Mimi's Variety, Paula and I were grabbed by a cop with a flashlight who'd caught us with beer. He marched us one block to the police station, where I was prepared to spend the night in jail for illegal possession of alcohol, but then the chief of police walked into the receiving area, recognized me from the gas station, and said, "She's a good kid. Let them go."

Another night I was pulled over driving my mother's car filled with six girls drinking and smoking pot. The cop asked for my license, shined his flashlight in my face. "You're the girl who works at Henry's," he said, then let me off with a warning, saved from DUI charges, possession of marijuana. Another night I was pulled over on Main Street not ten minutes after finding someone to buy us a case of beer. The cop — a different one from before — recognized me from Henry's. He confiscated the beer and waved us on. A day later the cop pulled into the gas station, and he and Henry stood by his cruiser talking loudly about the beer he'd enjoyed, smirking at me.

Sundays at the gas station were slow, except for people on their way to Walpole Prison, driving older-model low-riding sedans with rust or patches of Bondo, Cadillacs and Thunderbirds, dented and dinged, with holes in their mufflers coughing exhaust. The visitors, usually from Boston, were sometimes white, with a patina of toughness — bleached-blond hair, tattoos in an era when ink was not chic or middle-class — but most often they were African American, a stark indication of who lived behind the walls in Walpole. In Walpole Prison in the 1950s and '60s, the majority of prisoners were white, but the demographic began to shift in the 1970s as the war on drugs brought mandatory and harsher sentences, a crackdown that disproportionately targeted minorities. In the 1970s the "new" Walpole inmate was younger (under twenty-five) and black (35 percent), even though only 3 percent of Massachusetts residents were African American.

In the town of Walpole there was one black family. A couple of dozen black kids were bused from Boston to Walpole through the Metropolitan Council for Educational Opportunity (Metco), to attend the supposedly better suburban high school, as people in Boston battled over forced school integration. I glimpsed the busing riots on the news, kids my age hurling rocks at a school bus, someone's mother shouting, her face a rictus of rage. At Walpole High School, I never saw the Metco kids in the parking lot smoking or down the path getting high. They were athletes, club presidents, honor-roll students.

The Sunday drivers would buy gas, then ask, "How do you get to Walpole Prison?" I'd stoop to the window, point to the road ahead. "Go straight until you come to Route 1A in downtown Walpole, then take a left. In a couple miles when you see only trees, start looking for the prison on your right. You can't miss it."

One day a beat-up station wagon nosed up to the pumps with two large women filling the front seat, their thick pale arms dangling out the windows. They asked for directions to the prison and then wanted a few dollars' worth of gas. We were between shifts, and my coworker, Denise, was recording numbers on the gauges that tracked the volume of gas pumped. Denise, who wore her hair in a thick braid to her waist, was four years older than me, sloe-eyed, short like me, but with a commanding presence that made her seem taller. She chatted with me as she jotted down the figures, using the plastic price sign from atop the pump as a clipboard. The driver stuck her head out the window: "Hurry up!"

Denise muttered, "I'll take my sweet time" and then "asshole." The woman stepped out of the car. She was a head taller than Denise and seventy-five pounds heavier. She looked like a weight lifter or Roller Derby skater. "What'd you say, bitch?" I heard a car door creak and the woman's companion climbed out, and she was just as large — perhaps they were sisters, or a wrestling tag team. The driver swung at Denise and they fell to the ground. I stood frozen as Denise clutched a fistful of the woman's hair and bashed her on the head with the plastic price sign. Henry and the mechanics ran out of the station and pulled them apart, and the two women piled into their car and peeled away.

Henry took Denise into his office to comfort her, which may have been the moment they began their affair. They'd disappear into Henry's office for twenty minutes or so, leaving me to take care of the customers, Henry emerging with a Cheshire Cat grin. One day he said to me, "With her talent, she could be in the White House." He mimed a blow job, tongue in cheek literally. I'd read *The Washington Fringe Benefit,* by Elizabeth Ray, a thinly disguised roman à clef published that year about a congressman's secretary — named Elizabeth Ray in the "novel" — who sexually serviced men for favors. I shook my head at Henry's comment and walked away. It was *ridiculous,* I knew, even as some part of me wondered what secret power Denise had.

Denise needed a place to live, so Henry and his wife took her in, but a few weeks later they kicked her out. Henry said his wife found pot on Denise, but Denise told me the truth. Late one night when Denise and Henry were playing Monopoly, Henry reached across the table to fondle Denise's breast and his wife saw. Denise began to sleep at Fast Freddie's house, or his parents' house, a one-story ranch, which enabled Denise to climb through the window into Freddie's bedroom each night and out in the morning. Freddie was visibly ecstatic, as happy as a slow-moving, lethargic man could show. Now a woman entered his bedroom every night and fucked him, like a dream come true, a wet dream. Soon Freddie's parents found out and Denise was homeless again, but by then she'd quit the gas station and I lost track of her forever.

The whole strange world passed through the gas station: teenagers crammed into cars, townies driving beat-up shitboxes, businessmen late driving home from their commute along Route 128, bikers filling their tanks for three bucks, mothers with kids loose in the backseat. Nights were surreal, especially in summer. Cars were filled with people going places — lovers on dates, girls slid over close to boys behind the wheel, kids partying in full-sized club vans, the smell of pot drifting from the windows, the dull thump of bass muffling into the humid air. One night a man in a manual wheelchair rolled himself down the shoulder of Route 1, wheeled across four lanes into the gas station, filled his tires with air, then rolled himself onto the shoulder of the highway and disappeared into the night as if he were escaping from something or somewhere.

There was a sense of expectancy, of waiting for an accident to happen, knowing that someone somewhere would crash, that the phone would ring, the startling jangle amplified into the parking lot. Every so often at the busy intersection I heard the sickening crunch of clashing metal. I registered the impact in my body like shock waves, an almost physical reverberation. I never grew accustomed to the jarring collision sounds, which signified more than an accident; it was the sound of fate, the sound of someone's life being abruptly altered, something ruined. One night a car hit a Winnebago, and bathing suits and float toys and lawn chairs were strewn all over the highway, the image of a halcyon lakeside vacation mingled with glass shards and twisted chrome. Hourly — sometimes more often — the wrecker pulled into the lot hauling a dented or smashed car or one that had broken down, the station a repository for mechanical failure, for the consequences of human error.


My job kept me from smoking angel dust from 3 p.m. to 10 p.m. on weekdays, but it also gave me money to buy it. Sometimes I spent most of my $80 weekly paycheck on dust. Still, there was a net gain — I had less idle time to fill with the nothingness of a dust high.

A few months after I started work, Paula stopped dealing dust, at least temporarily, after her supplier, her brother Duane, totaled his car. Attempting to round a curve at two in the morning, probably high on dust or drunk, Duane slammed head-on into a giant oak tree in someone's front yard. The impact pushed the engine three feet into the body of the car, and Duane was rushed to the hospital. The day after the accident, Paula and I drove to the junkyard to see Duane's car, the car in which I'd first smoked angel dust a year and a half earlier, when the floor of Friendly's had turned spongy — the illusion of a soft landing.

As we walked among the dented cars, we saw ahead of us the rear of Duane's sedan. Closer, we saw the top nearly sheared off, peeled back like a scalp, and then we saw the engine thrust into the front seat, where Duane had been sitting, the half-moon indent in the grille, the shape of the tree. It was a horrific wreck, the accordioned metal testimony to the miracle that Paula's brother had survived. Paula cried, and I felt my throat close. Seeing the car was more visceral than hearing about the accident, knowing that Duane was lying unconscious in intensive care, bandaged up, big strong powerful Duane, who seemed so cool, so invincible, like a superhero of some kind — Dust Man. But now we saw two tons of steel crumpled like tissue. Bones broken, teeth knocked out, organs pierced, a body left nearly dead.

After Paula stopped dealing, something began to shift between us, and between me and Alison, or maybe something shifted in me. Affection for the drug became greater than our affection for each other. The drugs became so valuable, like gold, like money, and we suspected each other of cheating. "Me and Paula split a gram of dust with Alison," I wrote in my diary. "Hers looked twice the size of ours." I felt the shift, like ice breaking up on a pond and I was drifting on my own floe. Maybe I was not as cool as Paula or Alison. Maybe I was too serious, too morose. I started to feel that my friends included me because I had a car, or because they could party and crash at my house when my mother was away. Outside of getting high, we had little conversation anymore.

After being out with my friends one night, I felt desperately sad. I knocked on the door of Sally's bedroom, a partially refinished room in the basement that she'd taken over in her senior year. I could hear the television and knew she was in there with her boyfriend, Kevin, a happy-go-lucky boy who adored her. I sat on a chair in her room and broke down crying. "I hate my life," I said. "I hate my friends."

"You don't need them," Sally said. I was grateful that Kevin didn't look at me like I was an idiot, weeping in an embarrassingly sloppy way. I didn't really hate Paula and Alison, but I didn't like who I was when I was with them, who we'd become together. This was the first time that I had admitted to anyone that I felt bad, spoke aloud that truth. I still hung around with Paula and Alison, even though I suspected that they didn't like me, because it was preferable, somehow, to staying home alone. I'd lost the ability to keep myself company, or I was afraid that alone I would find myself in poor company.

One night as we drove around smoking angel dust — I was driving, Alison sitting shotgun, Paula and Nicky in the backseat — I looked in the rearview mirror and saw Paula and Nicky kissing. I was dusted so I couldn't make sense of it, how incongruous it was. *Is that what I saw?* After I dropped off my friends, downstairs in the Orange Room, Nicky said he wanted to break up. He still loved me, he said. "It's not that." We'd been together for a year and a half, but only five months had passed since he'd given me the diamond-chip pre-engagement ring.

Nicky slept over that night, as if nothing had changed. In the morning I drove him to his friend Tony's house, and perhaps out of guilt he invited me inside. In the basement a half-dozen boys from his neighborhood sat in a circle passing a joint — Tony, with his rippled lopsided hair, Mickey Flynn, whom I never heard speak a single word. I took a hit off the joint and passed it, but suddenly I felt too high. I stood abruptly. "I gotta go."

I slid behind the wheel of my car, which felt like a hovercraft, a boat floating on rolling seas, the car body rising and falling on swells. I concentrated on staying between the lines, gripping the steering wheel, driving slowly. I had to focus, figure out which turn to take, how to find my way home. I came to a jolting stop in my driveway, relieved, and I walked upstairs and crawled into bed, feeling stoned and dusted at once. Everything was distorted and strange and I felt far away from myself and sadder than I'd ever felt, as if I'd never not feel sad again, as if I'd never get out of bed, like Janis Joplin said, *tomorrow never happens, it's all the same fucking day, man.*

I stayed in bed all day, crying or staring without speaking, immobile, curled like a pill bug under the blanket. One by one people came into my room to talk to me, try to shake me loose from this near-catatonic state. Sally sat in the chair in the dormer nook, and as she spoke her face seemed to distort, then return to normal. I couldn't concentrate on what she was saying. Later that afternoon my father stood near my bed and spoke to me about college. I'd never felt so disconnected from him. I wished my father would stop talking about college, which I never thought about, which no one had ever mentioned before. Not knowing what to say, he reached so far into the future that I couldn't follow. He seemed as lost as I was.

I don't recall my mother talking to me that day when she returned from New York; she didn't have patience for a kid lying about, feeling sorry for herself. When Patrick failed to get out of bed for school, she'd dump a pitcher of cold water on his head. My mother and I rarely had mother-daughter "chats." She was too busy, worked full-time, had six other kids. Mostly she yelled, threatened to "lay you out in lavender," occasionally threw something and then moved on, the action-figure mother. I know she cared — everything she did was for us — but she left me alone, or trusted me, perhaps, to sort this out myself.

When I didn't come down for dinner, someone called Sue in her dorm, or maybe Sue called home and someone told her that Nicky had broken up with me. Someone carried the phone to me and Sue asked me if I wanted her to come home, and I said yes — the only word I'd spoken all day. Knowing that she was coming home helped somehow, because by the time she arrived two hours later I didn't feel so profoundly despairing. I was rising in my bathysphere from the depths into ordinary sadness. Sue didn't stay long, just an hour or so, and then she drove two hours back to her dorm, but somehow she'd broken the spell.

I have a Polaroid of Nicky and me in my living room, sitting on the scratchy floral couch, Nicky wearing his ever-present ski cap, plaid flannel shirt, hair to his shoulders. He looks straight into the camera, his face perfectly balanced, his skin porcelain, like in a Renaissance portrait, but he's not smiling. I lean into him, my head resting on his shoulder, my arm looped through his, as if he'll float away if I don't clutch on to him, or I will, like he is ballast.

At the prom that spring, our junior year, he'd sat sullen at our table, stoned on something, maybe cocaine, maybe dust, possibly both, me in my candy-apple-red gown with white confetti-sized polka dots, Nicky in a white jacket and pants, black cummerbund and red bow tie, a scarlet boutonnière. In a snapshot I took, he looks woozy and unhappy, and afterward when a bunch of kids crashed in the Orange Room, Nicky sat on the cellar stairs for hours talking to Maggie Brenner. I suppose our breakup was like my parents' divorce, which I didn't see coming either, since there were no fights between us, no trouble, just a slow drifting away from each other.

Nicky was a beautiful boy who loved me at a time when I didn't think anyone else did, not even myself. The sorrow of losing him tapped into a larger reservoir of sadness, but from that well some semblance of sense rose up, or perhaps just desperation. I asked my mother if I could see a counselor, and she arranged for me to see Jim, who'd been my parents' marriage counselor. On Thursday nights I left work for my appointment at 7 p.m. Henry found someone to cover my hours at the gas station. That first session, I parked in front of Jim's split-level ranch, walked up the flagstone path, up the brick steps with the wrought iron railing, and knocked on the door. Jim, a short, balding man with dark hair and a Grecian nose, answered. His house was pin-drop quiet. I followed Jim to a small room and sat in an armchair in front of his desk, as if I'd been summoned to the principal's office.

After we exchanged pleasantries, he paused, then said, "How do you feel about your parents' divorce?" The question caught me off-guard. I thought I was there to talk about *me*. But my response surprised me, too. I couldn't speak, and I began to cry, and then to sob. Jim waited, said nothing, waited longer. He nudged a box of tissue to the edge of his desk, then sat silently as I cried. He waited and waited, patiently. That night in Jim's office was the first time since the family meeting four years earlier when my parents announced they were separating that anyone had asked me how I felt about it. I hadn't known that my parents splitting up had hurt, that our broken family — broken home, as Mr. Gurkin called it — had broken my heart. Or I must have known intuitively, because my response was to throw up barriers to feelings, suppress pain, numb myself to it. "I don't care," I'd told my friends in seventh grade, and then spent the next four years making myself unable to care, or feel.

I hadn't known that our big happy family's falling apart was the source of the big huge cry. For years I'd harbored a fugitive sadness. "It will all come out someday though," I'd written in my diary. I apologized to Jim for crying and then cried more, embarrassed that I couldn't control this gushing, but whatever soothing words Jim said made my outburst all right. I smiled and cried more and finally it was time to leave. Opening that stuck spigot, allowing myself that big huge cry — that was all that Jim and I accomplished our first night.

I paid Jim with a $20 bill as if it were a drug deal, under the table, as if there were something slightly illicit about the interaction. Each week after that, I paid Jim with a $20 bill, money I earned pumping gas, a quarter of my weekly pay. I insisted to my mother that I pay for my counseling, perhaps as a way of paying for my sins, compensating for damage and destruction, the harm I'd caused a lot of people, including myself. I must pay.

# 8

## *Speech Acts*

TOM PARISI, THE NIGHTTIME TOW-TRUCK DRIVER AT THE GAS STA-
tion, was charged with some minor crime, so he fled to California, out-
running the local police. Suddenly Gerry, my boss's brother, drove the
wrecker at night. Gerry had gotten into some kind of trouble — drugs,
maybe; he wasn't the violent type — and had dropped out of college. He
was better-looking than Henry, taller, lanky where Henry was stout,
but with the same copper-brown eyes and long dark eyelashes, a fine
straight nose, a square jaw with a cleft chin, and perfect white teeth.
Gerry's hair was dark brown, shaggy to his shoulders. Compared to
Henry, there was something soft about Gerry, perhaps just youth, and
an impish look, perhaps the slight hint of yellow in his irises.

I came to know Gerry on sultry summer nights as we smoked cigarettes in the gas station lobby, waiting for an accident to happen. "Ask
me the capital of any country," he said one night, impressing me with
rote knowledge that I took for worldliness. He was twenty-one, and
I'd just turned seventeen that summer after my junior year. Gerry said
he was a voracious reader. I nodded. "I read the toothpaste tube in the
bathroom," I said. He smiled, and together we recited from memory —
*Crest has been shown to be an effective decay-preventive dentifrice* — and
that united us as soul mates, at least in my heart.

Gerry and I talked about books I'd read — *Stranger in a Strange Land,
Brave New World, Siddhartha, Anthem, The Stranger.* These books appealed to me with their abstractness, the language of metaphor, which

I understood easily, like Spanish; I could make that translation. Gerry gave me *The Teachings of Don Juan: A Yaqui Way of Knowledge,* by Carlos Castaneda, an anthropology student who'd studied shamanism in Mexico. One story in the book intrigued me. The shaman, don Juan, told Castaneda that there was a spot in the room that was his perfect spot, that when he found the spot, energy and strength would flow through him. For hours Castaneda sat and meditated, then he moved and sat and moved again, trying to find his spot, the spot where he'd feel harmony with the universe, the place where he fit. I wanted to find my spot, not in a room but in the world, the place where I was strong, the place I belonged.

Castaneda ate peyote and hallucinated that he was a crow soaring above the landscape, like the flying-mattress sensation I had when I went to bed high on dust. The book made sense to me: with enhanced perception one might find truth, like on the night earlier that summer when I'd dropped acid. It was one of those placid evenings in late June when daylight lingered so long that it felt like time had stopped, or was caught, the air infused with tension. Someone drove up to a crowd of kids milling around Mimi's Variety downtown with a sheet of blotter acid, rows of pinkie-sized Mickey Mouses stamped on paper, and for $2 a hit, a couple dozen kids tripped, just something to do on an empty summer night.

Paula and I swallowed our tabs and then hitched a ride to the high school parking lot, where kids were loitering, rock music blaring from someone's car speakers. Nicky was there, but he kept a distance. We'd barely spoken in the weeks since our breakup. Alison was somewhere else, with some boyfriend, or in Westwood Lodge again, or maybe another place, like the scary-looking halfway house near Boston where we visited her once — no plush Victorian furniture, just industrial bunk beds with metal frames in spare cell-like rooms.

Under the green-tinted sodium lights of the school parking lot, all the world's a stage. In my hallucinatory state, everyone around me transformed into caricatures: the tomboy, the dumb blonde, the football star. Their voices grated; their laughter seemed forced. The mood was strange and latent, like the silent static of heat lightning. I had a surreal sense of being outside of myself. *I don't belong here,* I thought. *I have to get out of here.* I leaned against a car and observed everyone as if I were invisible, as if I'd dropped through a hole in the earth, like in that grade-school book that I loved, *The Forgotten Door,* about a boy who fell through a mysterious tunnel in the ground and plummeted through space. He landed in a place where he resembled everyone but was different somehow; people and objects and the landscape were *familiar* but not the same, an approximation. Faced with this horrible realization, he wondered, *Who am I?*

This was the question I wrote in tiny print across lined paper during geometry, stoned and dreamy as Mrs. Drane's dreary words circled down a drain, never entering my mind. *Who am I?* The question returned to me like a boomerang, without an answer. But the boy who fell to earth had one strength — *he could read people's minds.* The Mickey Mouse blotter acid had a similar effect on me: I could see through people, see behind their masks. They were not real; none of this was real.

Later that night I said to Paula, "You can tell who your friends are because they can look you in the eye." I didn't mean this as a test, just something I'd observed, but when I said that, she looked away, which surprised me, and then everything felt awkward between us. The next morning I couldn't bring myself to call Paula, as I had every day for two years, and she didn't call me, and we never called each other again, and for a while that summer I had no friends. Except Gerry.

Gerry wanted me to meet his friend Quentin, who lived in Quincy, south of Boston. After we closed the station at ten o'clock one night, we drove up Route 1 in the Egg. On the way to Quincy, Gerry told me about Quentin, whom he'd met in college but who, like Gerry, had dropped out. "He works as a janitor," Gerry said. "He's mildly schizophrenic." Quentin from Quincy crazy with schizophrenia — the *z* in *schizophrenic* buzzed like the *z* in *crazy,* like the *z* in the razor blades that Alison used to cut her skin, like the word *crazy* felt like a razor blade in my mouth.

I'd read about schizophrenia in *I Never Promised You a Rose Garden*, a book that terrified me. The girl in the book slipped into a scary world of her own invention, the Kingdom of Yr, with gods who turned against her. The made-for-TV movie came out that summer. In one scene the girl sits in an office talking to her psychiatrist — a setting like Jim's office, a desk, armchair, paneled walls — when suddenly metal bars crash down around her, caging her. At least *she* sees the bars: a hallucination, only without drugs. But I saw the bars, too, in the film, and heard the menacing growling voices of the gods as if I were inside her head, a frightening place. The book was based on a true story about a real girl who was my age and who saw a shrink, like I did.

But Quentin was only *mildly* schizophrenic, Gerry said, without further explanation about the degrees of insanity. "He's a really nice guy." I gathered I was to be gentle with Quentin. I was mildly excited about meeting a mild schizophrenic, someone interesting, someone who dwelled on the fringe of society, where it seemed to me all the intrigue was, the excitement. We parked on a city street and Gerry ran up a set of stairs to fetch Quentin, who was short with a full beard and a bush of dark-blond hair but who otherwise looked ordinary and not mildly schizophrenic. Quentin sat in the front passenger seat and I was relegated to the backseat of my own car. It seemed that giving Quentin the front seat was part of the special handling he required on account of his mild schizophrenia.

We drove around on unfamiliar roads, drinking beer and smoking pot, until we landed back at the gas station, where we were safe, like home base, like gools. Quentin and Gerry talked and I listened as they discussed Nietzsche and Sartre, things I didn't know yet. Gerry lit a joint of angel dust. I wasn't smoking dust much since I'd stopped hanging around with Paula and Alison, since I'd been in counseling with Jim, but I hadn't quit altogether. I leaned forward, wedged in the space between the bucket seats like a dog, to hear the conversation between Gerry and Quentin, who was now sitting in the driver's seat for some reason. The car smelled faintly of gasoline, because I hadn't changed out of my work clothes, and the air felt electrified with a weird energy that could spark if Quentin began to act crazy, which I thought he might at any moment.

I asked Quentin if I could touch his beard. "I've never felt a beard before," I said. Somehow Quentin didn't see my request as strange, maybe because he was mildly schizophrenic, and on top of that drunk and dusted. I touched his pelt of a beard, which was softer than I'd imagined it would be. "It's really soft," I said. I petted the beard, and Gerry leaned over and whispered in my ear, "Keep doing that, he likes it," as if Quentin were a wild animal I'd tamed. I kept petting the beard, but then I didn't know when to stop, which was just one of the many, many things I didn't know: when a seventeen-year-old girl should stop petting a mildly schizophrenic man's beard, at midnight, parked in a gas station.

Gerry and I drove north one Saturday for the Gerry Delaney tour of Salem, a coastal town north of Boston, famous for the witch trials of the seventeenth century, though geographically the "witches" were from the village of Salem, which is now Danvers, an adjacent town. Gerry had gone to college there for two years, but he'd quit abruptly, and the town seemed to haunt him. We cruised past the House of Seven Gables, dark and brooding, and then downtown, where Gerry showed me the witch store, Crow Haven Corner, which sold incense and potions and tarot cards. The store was owned by Laurie Cabot, a famous witch, whom I'd heard on WBCN, the hip Boston radio station I tuned into daily to hear astrological forecasts from the Cosmic Muffin. "There are still witches in Salem," Gerry said.

After showing me the witch store, Gerry drove by a bar where he'd been stabbed one night, he told me, though he didn't say why he'd gotten into an argument serious enough for someone to stick a knife in his gut. He described how he'd stumbled out of the bar and crawled into an alley — he showed me the alley — and had lain there bleeding until "the homosexuals" found him, took him to the hospital, then took him into their home and nursed him back to health. Gerry called his saviors "the homosexuals," a formal term meant to show respect. We drove to their apartment next, on the third floor of a Victorian house. Gerry paused to look up, remembering the scene, the moment of his near-death, his rescue. We climbed a set of creaky wooden fire-escape stairs and walked into a small kitchen jammed with people.

While Gerry caught up with the homosexuals, I wandered into the living room, empty but for a lone woman in a black dress flowing to her ankles, swirling to the music, weaving a scarf through the air like an acid tracer, her movements slow and hypnotic, as if she were casting a spell. Gerry came up behind me, whispered, "She's a witch." Maybe the dancing woman was Laurie Cabot, the famous witch, or maybe she was just an ordinary everyday witch, one of Cabot's protégés.

The homosexuals were moving, so everything in their apartment was for sale. "Just look around," one man told me. On the wall in the living room I saw a Mona Lisa print. I loved the Mona Lisa, her enigmatic closed-mouth smile; she was keeping herself secret, as I did, or maybe just hiding poor orthodontia, as I did. I asked one of the men how much for the poster. "A dollar," he said, and I rolled up the print. I found a shallow wooden bowl, also a dollar. "That was hand-carved by Indians," he said.

After the party Gerry and I drove to a park, drank some beer, smoked pot. We fell asleep in the Egg and woke to blinding daylight. Gerry drove to nearby Singing Beach, where every step was a note, or at least a squeak, the silica in the sand squelching underfoot. As I walked I heard *sink sink sink* or maybe *sing sing sing* against the clap of waves. At this hour the beach was all ours. At the base of a fortresslike rock, I saw a glint of gold in the sand and reached for it, excavating a bottle of André Cold Duck. I'd never had champagne before (or sparkling wine, as it were), so the buried treasure felt like kismet, meant for Gerry and me. We climbed to the top of the rock and stared across the sea, salt mist in the air, on my tongue. Gerry popped the cork and we passed the bottle back and forth, swigging, greeting the morning silently.

With Gerry, Salem was fantastic, like a chthonic theme park with witches and homosexuals and dark nights of stabbings and a heroic rescue, ancient artifacts from Indians, glittering treasure beneath our bare feet, champagne and the ocean in our mouths.

In counseling, Jim asked each week if I'd smoked angel dust. He never scolded or berated me, or demanded that I stop, or even preached its dangers, but by asking he forced me to confront my choices, to remember that I had a choice. After talking about my parents' divorce, we turned to the question that plagued me, that I wrote in those spiral-bound notebooks like liner notes to the libretto of my life: *Who am I?* Jim said I was suffering from an "identity crisis." So I didn't have a complex, as I'd suspected, but worse; it was a *crisis*. An identity crisis made perfect sense; I'd been incrementally erasing myself with angel dust, like shaking an Etch-A-Sketch screen. I'd spent so much energy concealing my real self, suppressing my feelings, pretending I didn't care about anything, that I'd lost whatever self I'd developed before my teenage years.

"How do I know who I am?" I asked Jim one night. "You get to know yourself by being yourself," he said, which seemed as mysterious as the poster on Miss West's wall: *As soon as you trust yourself, you will know how to live,* a quote from Goethe. I sat in the back in Miss West's algebra class, near the window that overlooked the woods, daydreaming and scribbling in my notebooks, pondering that quotation as if it were a koan — *Trust yourself and you'll know how to live.* Those words bothered me. What did it mean, *trust yourself*? I was supposed to be working algebraic equations, solving for $x$, but *I* was $x$, the unknown. At least Jim's words — "get to know yourself by being yourself" — reassured me that there was a *self* inside me to be; I just had to unearth her, like a long-lost favorite dress. I didn't have to invent her; she was there. For years I'd been inventing someone who was *not* me; no wonder I did not like that girl.

I always arrived anxiously early for my Thursday night appointments with Jim, so I'd park nearby at St. Timothy's Church, in the empty lot overlooking New Pond, and watch the lake sparkle diamonds as the sun dipped, listening to songs that ripped me open, like "It's All Behind You," by Andy Pratt:

*And maybe today you came in, ripped your fingernail on the door*
*Ran into the TV set and then you said something*
*That made everybody think you were really stupid.*

I'd felt that crippling self-consciousness for too long — self-doubt, self-loathing, eased with numbing drugs.

December 15, 1975 — I did the most embarrassing thing in Spanish. Mr. Gurkin told me to pull up the map so I could write on the board and I pulled too hard and it fell on the floor and made a huge noise. I was so embarrassed. Craig, Monica Lande, Keith W., all those cool kids were in there. They were probably saying to themselves, 'What a fucking clod.' I almost died.

The diary entry shows my unforgiving self-consciousness, my feeling uncool, as if coolness was all that mattered, as if a lack of coolness was lethal — I could nearly die from it. Acting a certain way to fit in, while simultaneously suppressing feelings, made me hyperbolically aware of my every move. It was exhausting, which is probably one reason I turned to drugs; angel dust erased the self and thus self-loathing. Whatever squelched my voice in childhood — soap, pepper, slaps, shame, punishment — with angel dust, I silenced myself.

I can't remember *not* being self-conscious, analyzing and thinking, interpreting events as I watched them unfold, a narrative stream in my head. In first grade Miss Hanson asked the class how to pronounce the letter *A*, then called on me. I knew the answer — "Aaaaay" — but I was embarrassed to make that awkward sound. Miss Hanson waited. "Maureen, I know you know the answer." She saw right through me, like the teacher on *Romper Room* looking through her magic mirror. But I waited her out, silent. Miss Hanson stared at me with pleading eyes, then called on Betsy Costello, who said, unhesitatingly, "Aaaaay," just like the sound in my head that I knew was correct. Even then, in that small way, I knew I'd failed myself.

My self-consciousness was coupled with a compulsion to speak — that fresh mouth — and these two opposing impulses created inside me an anxious simmering feeling, like I could blow. The propensity to speak combined with self-consciousness made me something of a blurter. One night at dinner, when we still had family dinners, in a rare moment of quiet, I said, "Mrs. LaFarge takes acupuncture." This news had been bubbling inside me all day, this piece of exotica. Everyone laughed, which was not what I'd expected. I'd expected someone to say something like, "Oh, how interesting! How did you learn that?" followed by a discussion of Mrs. LaFarge's health problems and acupuncture in general. Instead my statement became the family emblem of a ridiculous non sequitur. To me, the moment stands for that inner struggle I had as a girl — between a native outspokenness and a native shyness, an impulse toward courage and overpowering fear. I felt like the pushmi-pullyu from *Dr. Dolittle,* my favorite book in third grade, the llamalike creature with a head at each end, working at cross purposes, like having your foot on the gas and the brake at the same time.

I was melancholy and moody, awkward and self-conscious, but I began to see those traits as positive. If I was self-conscious, that meant I was at least *conscious,* not absent, as I was on dust. If I was moody, at least I had a mood, felt something. I wasn't numb. As I smoked angel dust less and less often, there was a clearing in the mental fog, through which I began to see the waste of it all — thinking about getting high, preparing to get high, being high, crashing from being high. In the end, it was just boring.

As I sat in the church parking lot those twilights before counseling, I'd crank the volume in the Egg, filling the car with Andy Pratt's haunting soprano, a sudden plummet to baritone — *All your fears are gone, gone, gone, gone, gone* — like a bell ringing, an alarm waking me up. There was another Andy Pratt song I played over and over, "Inside Me Wants Out": the authentic self that I'd buried, effaced, that I'd nearly destroyed — she wanted out.

• • •

With Gerry I began to relocate my interests, to know myself by being myself, as Jim had advised. Like Gerry, I was bookish and philosophical and analytical, a little neurotic, a little political, hungry for something I couldn't find in Walpole. Gerry had cracked open the world, shown me its molten core, Gerry my unconventional teacher, with whom I'd begun to fall in love.

One weekend we drove into Boston for a free concert at the Hatch Shell on the Esplanade, featuring Heart, "the Wilson sisters," Gerry called them, as if he knew them personally. I hadn't been to the Hatch Shell since the Cowsills concert with my family a decade earlier. Now with Gerry I had a flashbulb memory: our blanket spread on the lawn, standing on tiptoe to see the band way up front, sitting on my father's shoulders like it was the top of the world. But the Cowsills, America's wholesome family band, were no longer *happy happy happy* as they'd sung in 1968. The band split up in 1972, the year my parents separated, and by 1977, when Gerry took me to the Hatch Shell, some of the Cowsill kids were struggling with drugs and alcohol and depression. One brother was addicted to heroin.

The lawn was jammed that afternoon for Heart, the field a patchwork of blankets, girls in halter tops riding their boyfriend's shoulders, bare-chested boys perched in tree branches to see over the crowd, the whiff of pot. Gerry took my hand as we pushed through the sea of people, and for a moment, with my hand in his, I felt like his girlfriend. But Gerry mooned over the sexy Wilson sisters and jealousy bloomed in me, a brine in my mouth. The crowd cheered raucously as the blond sister strummed hard on her guitar, the sound percussive like the hoofbeats of wild horses, and the dark-haired sister screamed *Cra-a-a-a-zy on you,* her haunting, witchy tremolo, then a whisper, *Every time I think about it, I want to cry.*

I wanted Gerry to kiss me, to fuse the current of desire between us. Gerry called me "jailbait," which I thought was a synonym for a tease, a flirt. I didn't understand the possible legal implications if Gerry had sex with a seventeen-year-old. Massachusetts General Law, Chapter 272,

"Crimes Against Chastity, Morality, Decency and Good Order," Section 4: "Whoever induces any person under 18 years of age of chaste life to have unlawful sexual intercourse shall be punished by imprisonment in the state prison for not more than three years." If we'd had sex, if someone, like my parents, had enforced that law, Gerry could have wound up in jail, though I'm not sure I could have proved I was "of chaste life."

My mother found loose change on the floor of my car that summer and told Sue she was worried that I was having sex, that the coins had fallen out of my pockets during sex. My mother's fears were those from her adolescence, unwed pregnancy being the worst thing that could happen to a girl. But she misread the evidence. The change had fallen on the floor because I'd been fucked up, too high to be able to conduct fine motor movements, like opening a purse; too fucked up to fuck.

My mother needn't have worried because Gerry never acted on the erotic energy between us, the electric circuit when we sat close in the Egg and talked. Things between us became cross-wired. Feeling rejected, I pushed Gerry away, or I picked at him to demand his attention, to get under his skin, to release a charge, which only pushed him further away. This was the opposite of what I wanted, but that was my modus operandi since my father had left — if I felt rejected or unseen, unloved, I pretended I didn't want or need you, pretended I didn't care.

Gerry began to frequent the local dives, smoking angel dust more often — Walpole was flooded with it — just when I was trying to quit. He turned sullen and irritable and started to fuck up at work, and then he was arrested, some bar fight. Henry called in a favor and Gerry was released from jail. Gerry had come to Walpole to leave trouble behind, but in Walpole he found trouble anyway. At some point Henry became fed up and fired him, or Gerry quit. Whatever happened, one night Gerry no longer worked at the station. I don't know where Gerry went, but I never spoke to him again. Summer was gone, and with it my erstwhile mentor.

In my junior year I'd spent more time at the gas station than in school; I missed thirty-eight days of school and racked up fourteen times tardy.

Work *was* school. For pumping gas I earned five credits through the school's work-study program. The work-study teachers must have thought I'd learn something at the gas station, and I did, though not what they might have expected. The job was like the practical lab to my sociology class, "America in the Twentieth Century." The gas station was where I learned about sexual harassment (the Supreme Court recognized the term for the first time that year, 1977); how women traded sex for basic necessities, a place to live; homosexuality; witches and the occult; the socioeconomic and race demographics of prisons, based on visitors who passed through the station; the petty corruption of small-town cops (even though it worked in my favor); the country's dependence on oil from the Mideast, the embargo flaring up every now and then, creating long lines of impatient, anxious customers as I changed price signs daily, a few cents at a time, but the price doubled by the end of that year; the dull safe sameness of American life in a parade of cars guzzling their precious foreign fuel.

On the first day of school in my senior year, September 1977, I smoked angel dust for the first time in a month and the last time in my life. Some kids were in the parking lot smoking dust before the morning bell and I joined them. Maybe it was the anxiety I always felt on the first day of school. Maybe because school, for me, had been a locus of intoxication, high school a place where I went to get high, and so I fell back into the habit of the place. Perhaps because I hadn't smoked angel dust in a while and my tolerance was lower, after I got dusted that morning, gloom fell over me like a shroud. I felt an almost physical sadness, not just a broken heart but a broken body, but I couldn't cry. I never cried while dusted — the anesthetic properties of the drug prevented that.

The sky was gray — at least in memory — and I hallucinated a chain-link fence around the parking lot. I felt locked out and also somehow locked in. I couldn't tell if the chain-link fence was real, if it had been erected over the summer, or if I was imagining it. Whatever conjured the image, it gave rise to a sinking despair powerful and frightening enough to deter me from smoking angel dust ever again.


In school I reconnected with Terry Littlefield, a girl I'd known from gymnastics before I quit, and she introduced me to Angie Harper and Linda Kelly, and the four of us grew close, cooking spaghetti dinners and drinking cheap rosé, having heart-to-heart talks, practicing being adults. We went clubbing. The drinking age was eighteen, but we had fake IDs. I danced wildly under strobe lights on parquet floors in dive clubs in Walpole and just over the state line in Rhode Island, a club called the Edge, flinging myself around like a happy fool, sticky with sweat, my hair pinned up with cocktail straws. I spent myself on exuberant eurythmics, the joyful spontaneous physicality I loved as a kid twirling around the living room to my father's boogie-woogies. Dancing was the opposite of the ataxia of angel dust.

Ed gave me his old pickup truck, a 1966 Chevrolet with three speeds on the floor and a cab over the bed, its windows adorned with black-and-white gingham curtains like the ones my mother had sewn for the camper. I found some old couch cushions to line the truck's bed, and someone gave me a plastic sign that said PARTY, red letters on a white background, which I stuck in the window, as if inviting the cops to stop me. Sometimes I drove the truck to Friendly's parking lot and loaded up with anyone who was around and we drove to the Braintree Twin Drive-In. Sometimes a half-dozen kids piled in and we headed for Cape Cod, sleeping in the back of the truck in the parking lot of a beach, swimming at midnight in the warm salty sea.

When I quit dust, my consciousness awakened, and my conscience, too; I had to pass through guilt and shame, like wading across the River Styx back to the earth side. But on that shore was a luminous landscape, a brightness to life, a reverberant joy. Until I felt happy I hadn't understood how unhappy I'd been. I grabbed at pleasure as if making up for lost time — the bounty of beautiful boys to kiss, to lie with on a sleeping bag in an open field under stars, or in the back of my pickup truck, tucked under a blanket together in innocent sleep. Everything was interesting, my mind, my heart, wide open.

In school I was *present* as I hadn't been in those lost sophomore and junior years. I rediscovered my love of learning, of thinking, like remembering myself. In "Creative Expression" I wrote poems about Gerry Delaney and Salem, about Alison and Westwood Lodge. Mrs. Springer, who taught "Modern Poetry," took our class to the Charlwell House nursing home, where we sat in a circle with a dozen men and women in their seventies and eighties, writing poetry to prompts, then reading our words aloud. On our last visit, we celebrated with refreshments and someone put a record on the stereo, an old-time jazzy number. I extended my hands to an old woman to help her stand and she danced with me and then more of the men and women rose to dance. Afterward Mrs. Springer thanked me for bringing life to the party.

After working at the gas station for thirteen months, in January of my senior year I asked Henry for a raise. "Why should I give you a raise," he said, "when I can hire anyone off the street?" I mentioned my experience, that I'd never called in sick, that I was the only one who cleaned the restrooms. Henry refused to pay me another quarter per hour, so I quit. He was surprised, and maybe a touch pissed off, but I gave him a two-week notice, so he was free to hire anyone off the street. It felt empowering to assert my self-worth, even if that worth was just twenty-five cents.

That spring, after I hadn't smoked angel dust in six months, I thought about quitting counseling, too. Talking to Jim week after week for a year — this middle-aged man, a stranger who listened for twenty bucks an hour, the cost of two grams of dust — slowly pulled me back into the world. But lately in sessions I'd recite the events of the week and Jim would nod. He stopped offering advice, or if he did, his words were familiar. Intuitively I felt I was done with Jim, the way I was done with the gas station job, done with angel dust, done with Paula and Alison. That part of the journey was over, the stretch where I needed

Jim to guide me, the way I'd needed training wheels on my bike when I was five.

When I nervously told Jim I thought I'd stop counseling, which felt awkward, like breaking up with him, he said, "Do you think you have more work to do?" I nodded. I knew I did. "I'm sure someday I'll be back in counseling." In a strange way I felt that I'd *outgrown* Jim, that I'd reached the limit of his abilities. Maybe I'd just outgrown myself, that troubled girl, like shedding a skin.

That June I turned eighteen on the same day I graduated from high school, and I felt a kind of aching happiness. For my birthday/graduation gift, I asked for money to take voice lessons. Earlier that year my friend Angie and I had met some boys in a garage where their band was practicing. We were to be the female vocalists in their all-male band. That afternoon Angie held the mic and sang mostly in key to Eric Clapton, *If you want to hang out / you've gotta take her out, cocaine,* and *Yes, you look wonderful tonight.* The boys asked me to try, but I was mic-shy, or perhaps I understood that, like my father, I couldn't carry a tune.

I had no burning desire to be a singer. When I asked for voice lessons — not singing lessons, not music lessons, but *voice* lessons — I think that I wanted something more literal: I wanted my voice back. I wanted to learn how to speak my mind, how to refine my smart mouth, that superpower — Critical Girl, with her laser vision and razor tongue. I wanted to find the courage to voice my thoughts, to speak without embarrassment, without the crutch of alcohol or drugs, to reclaim what had come naturally to me but had been tamped down because I was too assertive, too opinionated, too demanding, too truthful, too critical. *Loudmouth. Back-talker. Fresh.*

My father gave me a check for $100, "voice lessons" noted on the memo line, but I never took lessons. I used the birthday/graduation money to buy cocaine and I blew it all in one night. As Jim had suspected in our last session, I had more work to do.

· · ·

For my birthday/graduation, my parents also gave me a sleeping bag and a camera. After a year of art school, Sally was moving to California with her friend Vickie. I was going along for the ride. Terry Littlefield, my friend from senior year, would fly to San Francisco and meet us for the final week of our trip. In August we said goodbye to my family and drove in Sally's used Datsun up our dead-end street, California bound, the hatchback of Sally's car so packed we couldn't see out the back window, even though Ed had built a storage box on top of the car. The cooler in the backseat was stocked with cold cuts from my uncle who owned a pub, logs of bologna and salami and blocks of cheese — we were setting off to see the country loaded with luncheon meats. We had a few errands before we hit the highway — bank, gas station, hardware store — and then it was noon and we were suddenly hungry, so we picnicked in the parking lot of a strip mall in Norwood, one town over, laughing at how we'd failed to get far in our journey.

Soon, though, we were cruising Interstate 90, that promising black strip of highway, driving and singing, looking out the window, Vickie and I smoking cigarettes, through New York and past Lake Erie where the topography changed, the flatness of the Midwest like nothing I'd ever seen. We passed through Ohio singing, *four dead in Ohio / gotta get down to it,* and into Michigan, *took me four days to hitchhike from Saginaw,* and then Indiana, where we exited in Gary for gas, the air sulfuric and oily and mustard-colored from billowing smokestacks, and we sang, *Indiana wants me / Lord, I can't go back there,* the singer on the lam from the law, like my former coworker Tom Parisi. We drove and drove until we hit the snarl of traffic outside Chicago and we changed our tune, *Daddy was a cop on the east side of Chicago / back in the U.S.A. back in the bad old days,* slipping into a song that was an invitation — *Won't you please come to Chicago just to sing?*

We were free and we were young and we were beautiful, though we barely knew it; this was our epic journey, this was us seeking the world, reaching for it, three teenage girls, one just turned eighteen, two nineteen, looking for something we wouldn't know until we found it, singing our way there, *happy happy happy.*

Somewhere in the Midwest our brakes started to grind. We noticed it first at a tollbooth. We pulled off the highway to a hole-in-the-wall garage, one bay and an antiquated gas pump, and the mechanic delivered the bad news: we needed a whole new set of brakes, $90, a huge chunk from our kitty. We had no credit card. I'm not sure what gave us the courage to decline his offer — perhaps my experience working near mechanics — but we sensed that this guy was sketchy.

On our way out of town we saw a brake-and-muffler franchise. We sat in the lobby as the mechanic, a guy our age, raised the Datsun on a lift. After a while he walked over to us, wiping his hands on a greasy rag. "There's nothing wrong with your brakes." We told him what the other mechanic had said, and he shook his head. "They look fine to me." Back on the interstate, we heard the awful grinding. Eventually we realized that the grinding occurred when we drove over grooved pavement, rumble strips, which we'd never seen in our small town of Walpole, our small state of Massachusetts. We dissected the incident for an hour — pissed at the asshole who'd tried to rip us off, congratulating ourselves on our cleverness in seeking a second opinion, thankful for the honesty of a young man, laughing at our naiveté, how stupid we'd been for not knowing about rumble strips.

In South Dakota we drove through the convoluted spires and gulches and geologic formations of the Badlands, a name that promised outlaws in ten-gallon hats behind every butte, gunslingers in spurred boots, bad men in a bad land. We dutifully photographed the touristy but still astonishing Mount Rushmore, then drove through the night bleary-eyed, through thunderstorms with curtains of rain — we could see only a few feet ahead — a trucker at 3 a.m. nosing our bumper, forcing us to speed up, then passing and cutting us off, playing dangerous highway cat-and-mouse, scaring us into pulling into a deserted rest area, hoping he wouldn't follow, then back on the highway into the downpour, one of us hunched over the wheel trying to see the lane through blurry blackness like being underwater, another one sitting shotgun to keep the driver awake against the hypnotic windshield wipers, the third curled up in the cramped backseat trying to sleep, driving for hours and hours, pushing to get there, to California, delirious with fatigue, *Lord, I'm five hundred miles away from home*, and then the sky lightened and I saw something in the distance, a formidable cloudbank, but as the sun at our backs lit the horizon, there before us were the Rocky Mountains, a great ragged wall of land, the snowcapped peaks rising into the heavens, breathtaking, like something out of a dream, grandiose mountains that belittled the adolescent peaks of New England.

At midnight, exhausted, we pulled up to the welcome station at Yellowstone, but the ranger said the campgrounds were full and turned us away. Where were we supposed to go after we'd driven hundreds of miles on steeply graded mountain roads? We drove a mile back in the direction we'd come and saw a dirt road, which we followed to its end. We set up our tent in pitch blackness and climbed in and fell asleep. In the morning when we peeked outside we saw the steep face of a mountain and at its base a glorious rushing river. We washed in the icy river, then drove into Yellowstone, stopping at thermal pools, deep holes with boiling turquoise water, at the bottom the blanched bones of some fallen creature. We waited faithfully for Old Faithful, and on schedule it geysered from the ground, shooting a plume of water a hundred feet, as if the earth were the back of an enormous whale.

We drove and drove—I never tired of looking out the window, watching the world like watching a film. We arrived at the Great Salt Lake at dawn and tried to sleep on the salt-stained beach flat as a mirror, but we were swarmed by small biting flies so we gave up, got in the car, and headed for Lake Tahoe. We set up our tent in a beautiful wooded campground a short walk from the sandy beach, football-sized pine cones littering the ground.

That night we met two brothers, one our age and one in his midtwenties wearing a Stetson, and we sat around a campfire drinking beer and talking. Around midnight Sally and I rose to leave, but Vickie said, "I'm staying." Sally and I looked at each other. It didn't seem like a good idea, but we had no control over Vickie, who wanted to make out with the cowboy-hat brother. The younger brother seemed to feel as awkward as we did with this sudden turn. He headed back to their hotel room as Sally and I walked through the woods and unzipped our tent and climbed into our sleeping bags, though I doubt either of us slept. Hours later, Vickie slipped into the tent.

In the morning Sally and I cooked breakfast, eggs and fried Spam — someone had given us cans of the stuff — then we cleaned our campsite, packed the tent, and organized the car while Vickie sat at the picnic table penning postcards to friends. Sally and I exchanged looks. This was the first sign of discord in our small traveling party.

In San Francisco we met Terry at the airport, then drove to Muir Woods. I pressed myself against an enormous redwood, my arms stretched to show scale, to feel with my whole Lilliputian body the monstrously massive, beautiful tree. We slept on Stinson Beach, then drove down the coast, navigating the winding Pacific Coast route, which dropped away steeply beyond the guardrail to desiccated scrub far below. It was Vickie's turn at the wheel, and as she took a curve she lit a cigarette, steering the car with her elbows, which freaked us out. Then she flicked the cigarette butt out the window, ignoring — or having missed completely — the frequent signs warning of $500 fines, the danger of starting a fire. Vickie was put off by our scolding — the mood had changed since the night in Tahoe when she went off with the cowboy-hat dude. Meanwhile, the storage box that Ed made slowly slid off its frame, slipping inch by inch, obscuring the rear window.

Mexico-bound, we stopped at Venice Beach, where Sally and I had smoked angel dust on that trip with my father three years earlier, though now I could see what I hadn't back then — that Venice was rundown and dirty, the boardwalk unpeopled, just a couple of home- less men in tattered sleeping bags and strung-out kids huddled under Mexican blankets, as if a rogue wave had washed everything away. We camped at San Onofre State Park, selected from our guidebook without realizing it was in the shadow of a nuclear power plant. There were fast-scurrying mice everywhere, zipping along and disappearing into burrows in the dunes, which made me think that some nuclear leak- age had caused a freak explosion in the mouse population. That night as we sat around our campfire, two men invited themselves over, just plunked down next to us, long-haired bikers in jackboots and jeans, bandannas around their matted hair. One of the bikers casually removed his prosthetic leg, propping it next to him. Always there were men who found us, followed us, some we had to shake.

We crossed the border into Tijuana, a sad dusty town with skinny barefoot boys peddling packs of chewing gum, collage-shacks on side streets, tourists bargaining in embarrassing pidgin Spanish with people who didn't have enough money for shoes or decent houses, trying to save what amounted to pennies on some trinket they didn't need because that's what the travel guide said you were supposed to do, barter with the locals.

After a week we drove to the Los Angeles airport so Terry and I could fly home, but Sally had come to California to live. I hated saying goodbye to my sister, leaving her there with little money, just the car and a suitcase and a friend with no common sense, just the two of them now in California to make their lives; they'd have to find jobs, a place to live, wake up each day in a strange land where they knew no one but each other. I sobbed as I watched Sally drive away, sick with worry.

Terry and I signed up for a stand-by flight and waited for our names to be called, sprawled on the airport floor for twelve hours until finally we were on a plane headed east. I was weighted down with my backpack and a manzanita-wood walking stick I'd found on Stinson Beach, its surface engraved with trails from worms, like the contour lines on a topographic map etched into its pulp. I carried this stick on the plane from Los Angeles to New York City, on the train from New York to Boston, on a bus from Boston to Framingham, where someone picked us up and drove us home to Walpole. How many strangers had I accidentally jabbed or tripped with that walking stick?

The trip was just three weeks, but I'd driven as far away from home as I could in the contiguous United States, passed through the vast pancake plain of the Midwest, up and through the sky-scraping mountains, crossed the continent and stepped in California, then returned to home base like touching gools.

• • •

I didn't go to college that fall, even though — surprisingly, given my grades — I'd been accepted at the University of Massachusetts. I wanted to take a year off to save money, to figure out what I wanted to do with my life. I had no idea how to chart a course for the future beyond the vague "aim" I'd recorded in my high school yearbook: "to finish something I start," an acknowledgment of my history of quitting, giving up what I loved, giving up on myself.

One Saturday in July when I was ten, my father took Sally and me to the Lions Club track meet at the high school athletic field. I entered the 440-yard relay race — once around the track — with three girls from my school. Patty Lewis shot out fast and we led, but then she handed off to Sheila Barton, who ran awkwardly, as if her legs were tangled; the way she ran, her gait circular, reminded me of my mother's manual egg-beater. I waited anxiously with my hand stretched behind me, wincing as runner after runner passed me and advanced down the third quarter of the track.

Finally Sheila handed me the baton and I ran with all my might in my curved lane, passing one runner and then another, but there were still two ahead when I handed the baton to Regina Richardson, the fastest girl in our grade, but even Regina could only pull us into third place, a white ribbon. My face was red from effort, my hair matted with sweat, but my father was there at the finish line, smiling and proud that I'd tried so hard, and that's all I wanted and needed. I wanted to try that hard again, to make my father proud of me, to be proud of myself.


I worked three jobs that fall, which kept me busy, out of the house by 8:30 a.m. and home at midnight. I cleaned a doctor's office one night a week, scrubbed three toilets and seven sinks, dusted and vacuumed and emptied trash. I worked as a cashier at Stop & Shop supermarket and as a short-order cook and dishwasher in the cafeteria at the Bird & Son factory in East Walpole, down the street from Bird Park, where my father had taken us to pick chestnuts, the park a gift in 1927 from the Bird family to their workers and the people of Walpole.

For a century Bird & Son earned huge profits selling asphalt and asbestos shingles, running three shifts to keep up with demand, but during the oil embargo in the 1970s profits suffered. By the time I worked there, only two shifts operated. I drove through the gates of the chain-link fence surrounding the compound — I recall barbed wire, but I'm probably conflating the factory on the east side of town with the prison on the west side — and down an asphalt plane, like a sloping lawn of pavement, past four-story brick buildings with huge windows, their tiny panes opaqued. The buildings were connected in a *U* shape, with the cafeteria, a single-story cottage, tucked in the center. There was one way in and one way out, like the street I lived on, a dead end.

The cafeteria was run by Frank DeRose, a jolly, short, fat Italian man with bushy white eyebrows circumflexing rheumy brown eyes. At the front of the cafeteria were steam tables to warm pots of soup and vats of Frank's homemade meatballs, stainless-steel pans filled with ziti or sausage and peppers. A glass shelf showcased slices of pie and cake, and at the end of the serving line was an old-fashioned punch-key register. Behind the alley was a stainless-steel refrigerator, a butcher block, a grill, its steel vent-hood brown with grease, and the Bunn coffee machine, the profit center of the enterprise, where at twenty-five cents a cup Frank made his bones. In the small back room there was an oven, an industrial dishwasher, sinks, cupboards, a bathroom.

I worked alone from 7 p.m. to midnight, some weekends, and the busy day shift in summer. In the kitchen, Frank cooked pans of his delicious meatballs filled with chunks of leftover Parmesan, and giant pans of chicken cacciatore. He'd splash red wine into the tray, then swig from the bottle, singing or humming, bustling around the cafeteria, yelling greetings to the men: "Howie, what can we do for ya, babe?" Often I saw Frank peel a twenty from the thick roll in his pocket for one of the men who'd run into financial trouble. Frank was quick to laugh, guffaw even, wiping tears from his eyes. I loved watching him wring joy from any ordinary moment in that cramped factory.

• • •

In the back room I loaded china plates and cups into the stainless-steel washer, steam billowing out when I opened the hatch, a blast of wet heat on my face. I poured coffee and worked the cash register and, when I was alone, cooked hamburgers, or steak and cheese sandwiches, omelets — everything flavored with a hint of cigarette smoke. During the busy lunch hour, three of us worked the line. My coworker Gert was a woman of enormous proportions. She wore flat open-toed mule slippers and stretchy nylon pants. She had a big round face and a big round nose, her skin coated with liquid makeup, and short dyed-cinnamon hair that she curled with curlers. Throughout the morning she reapplied orange lipstick and flirted with the men. I was in awe of Gert's self-esteem.

Missy, my other coworker, was a waif with baby-fine blond hair and a vulpine face. She wore tight, low-cut shirts that revealed her cleavage when she leaned across the counter. She and Gert were quickly buddies, excluding me from their frequent conversations about sex, which was a relief. They sucked long drags off their cigarettes, Missy blowing a stream of smoke out the side of her mouth, Gert letting the smoke seep out her nose while she talked. I couldn't help but see Gert and Missy as those cartoon characters, Peter Potamus, a purple hippo, and his tiny monkey sidekick, So-So. One day I came to work and Missy was gone. Fired. Frank told me he'd caught her stealing from the till.

Frank shopped for supplies at Stop & Shop, and once in a while he checked out at my register. I sensed that he wanted me to pass some items through without ringing them up. I'd done this for Nicky's mother, who didn't have a lot of money. Seeing Frank in Stop & Shop with his boxes of day-old Entenmann's cakes marked down, which he quartered and plated to sell for a buck apiece, his "manager's special" packages of hamburger, cheese ends from the deli, I wanted to give him a break, but this would have demonstrated that I was dishonest, stealing from my employer. How could he, my employer, trust me if he saw me stealing from my other employer? I opted for honesty. I didn't want to lose what I valued most, Frank's respect. Perhaps, too, I didn't want to lose something I was barely aware that I was developing: self-respect.

Men filed into the cafeteria for lunch or dinner, or pie and coffee, depending on which shift or break, always coated in grayish white dust, ashen like half-ghosts, dust thick in the creases of their necks. To make shingles, felt paperboard was coated with liquid tar, then dusted with mineral grit. Every third man was missing a finger from the cutters, often an index finger, or the middle, sometimes both. It was jarring at first as they picked up their coffee cups with their thumb and middle fingers, the index stub pressed against the cup for stability, like a tripod. These were the lifers at Bird & Son; they talked of camping and fishing, cars, the track, conversations I overheard as I collected dirty dishes and wiped down tables. They called me "hon" or "darlin'" as they flipped a quarter onto the stainless-steel counter, happily balancing their hot coffee, black or light and sweet.

There was a contingency of young men, none of whom planned on staying. Harry was probably three years older than me, with stringy brown hair in a messy ponytail, globs of tar stuck in his hair, a single front tooth missing. He was often stoned at work. I could see why someone would need to be drugged to endure eight- or ten-hour shifts there, five or six days a week, how a stultifying assembly-line job could annihilate you from the inside out. Harry walked into the cafeteria one day with a bandage on his hand, a fresh injury, a finger sacrificed to the cutting machines like paying dues, propitiating the gods of corporate profit. If you worked long enough at Bird & Son, you paid in flesh and bone. Some of the young drugged-out factory workers like Harry "spilled" hot tar on their skin to collect the $50-per-inch insurance payout, less if it wasn't a third-degree burn. Easy money, I suppose. Skin grew back, unlike fingers.

Working at Bird & Son was a tour through factory life, like the tram ride at Disneyland, men toiling, trading their strength, their bodies, for dollars per hour, brawn the currency, as it had been for my grandparents on both sides — housekeeper, cafeteria worker, forklift driver, groundsman — the life my father had lifted himself out of through education, recognized at eleven by a priest as "exemplary" and given a scholarship to prestigious Boston College High School. My father worked in factories in high school and college — a wire factory and a sausage factory, both in the city — but education was his ticket to the middle class, like so many children of immigrants and their children, like me, as soon as I figured out how to get myself to college.

Work became a structure for my life, like a Zen practice — chop wood, carry water — only it was cook omelets, pour coffee, or ring groceries, bag them, or clean toilets, empty trash. Work helped me stay on track, to eschew hedonism for discipline. I was good at work. Work was something I understood from my earliest paid labor, mating socks when I was six and seven. Every day in our house when I was growing up, nine people's socks fluttered down the laundry chute to the cellar, 126 dirty socks per week, 546 socks a month, 6,552 socks each year: endless socks. There was always a laundry basket of unmatched socks and I'd patiently mate them, piecework for which my mother paid a penny a pair. I liked the meditative task, the sense of accomplishment, the rich rattle of nickels and dimes in my Band-Aid-box bank, a coin slot stabbed with a butter knife into the soft tin.

My mother worried that I'd never get to college, so that December she issued an ultimatum: enroll in college or move out of the house. She wanted me to have the opportunity of an education, as she had not. In high school my mother was vice president of her senior class, president of the student council, president of Theta Sigma Tau sorority. For a yearbook story in which her classmates imagined the future, my mother was cast as the first woman president of the United States, President Starr. She dreamed of attending college, but the guidance counselor told her she should get a job and help her widowed mother and younger brother. Later, in 1960, when she was twenty-two and had just three babies, she wanted to enroll in college, but my father thought that the family would suffer, so my mother surrendered her dream again.

Go to college or move out — my mother was serious. I hadn't a clue how to find and furnish an apartment, so I enrolled at the University of Massachusetts at Amherst in the middle of the academic year. There was a benefit to having misspent my high school years. After a while I grew bored with the UMass ("ZooMass") party scene, all those strait-laced kids going crazy with their first taste of freedom from parents. Still, I had no idea how to be a college student. I'd loved my classes in my senior year of high school — creative writing, poetry, psychology — but I'd never done homework, never studied. After my first semester at UMass, I was placed on academic probation, an echo of my junior year, when I'd nearly failed. That threat, the idea that I would be kicked out of college, motivated me.

I chose interpersonal communication as my major, with the vague notion of becoming a family counselor to help families like mine after they'd fallen apart; I wanted to glue together all those broken homes. In classes I studied language patterns and speech contexts and persua-sion theory and the dynamics of small-group communication, and as the classes became more theoretical, I couldn't see how I was going to help any messed-up broken families with dry theories on "speech acts."

In my junior year I took a creative writing class. Sitting cross-legged in someone's living room, reading aloud short stories and poems, I felt like I'd found my spot, the way Carlos Castaneda had in *The Teachings of Don Juan,* the place where I was strong, where I belonged. The sto-ries I wrote for workshops were autobiographical or biographical, sto-ries from Walpole, stories I couldn't shake from my mind, like the story about my dental hygienist, Valerie Ray, Sally's classmate, who'd been charged with forgery, larceny, and arson. She'd set fire to the dentist's office to destroy evidence of her embezzlement. Because of my "twisty teeth," I'd spent many hours in that dental chair with the doctor work-ing in my mouth, Valerie beside him. I wrote Valerie's story in the first-person point of view because it was easy to occupy the consciousness of a young woman committing a crime.

I found kindred spirits in writers and artists and musicians and activists, like my friend Beth, a gifted painter with strawberry-blond hair, pale blue eyes, and a faint mulberry birthmark on her cheek, like the fingerprint of God, and Becky from my dorm, who recruited me into the women's issues team and the antiracism team of the student government, where I gave a presentation on the sterilization abuse of American Indian women. I researched the Nestlé Company's infant formula campaign in underdeveloped nations, which caused babies to become malnourished, spending hours in the library's musky basement, where government documents were housed, using my critical skills for something worthwhile, kindling my passion for politics, for justice.

My father passed to me his *Mother Jones* and *Nation* magazines, with passages he'd aggressively circled and annotated in his barely legible handwriting. I remember how when he'd lived with us, he'd scowl at Nixon on television, call him a "bum," the worst insult I heard my father say. Politics and books became the bridge to reconnect with my father, the start of our lifelong conversation.

Sometimes I wonder how I survived those high school years, how I wasn't maimed or killed the hundreds of times I got into cars with people so fucked up they could barely stay in the lane, or by getting behind the wheel of a car myself after drinking, smoking pot or angel dust, taking speed or acid. Worse, I might have hurt or killed someone else. I could have been raped by strangers with whom I hitched a ride, or by boys on whose couches or in whose cars I passed out.

I know girls who weren't as lucky, at least the ones who talked about it anyway, and I know the boys who raped them, hometown boys; one was a kid who sold me a joint of dust, noted in my diary on December 1, 1975; one was the older brother of a boy I liked, who attended the junior high across town, from one of the most respected families in Walpole; the third was an older townie with California-surfer good looks, who one day at a pond laughed at me when I said I could swim across. *Not only can I, but I'll beat you.* In the water I took off, then floated on my back as I watched him far behind, paddling clumsily like a dog. I found out later that he'd sexually assaulted my friend's sister while she was passed out at a party.

I could have become a burnout, a space cadet, circuits blown, a no-mind. I could have died of an overdose a hundred times if I hadn't passed out first, or run out of drugs, or been prevented in some way from ending up in oblivion because I did not know how to stop. I could not stop.

In junior high, sometimes when my father visited I'd shoot baskets in the hoop at the top of our driveway and he'd wager a quarter that I could sink the ball from what might have been the three-point line. When the ball swished through the net, he'd say, "Double or nothing," raising the stakes once to $32 as I hit the sweet spot again and again. But I always went too far. He knew I'd keep trying to impress him, keep shooting for more and more money, but he also knew that I'd miss eventually. I never quit while I was ahead, before I lost everything.

When I consider what could have happened, I become momentarily unhinged from my present life, flush with a loosed anxiety that somehow I might still be in danger, or dangerous. If I was that reckless with my life, that capable of stupid, careless decisions, if I was senseless then, I could be senseless now — untuned to the danger signals around me. I feel so narrowly escaped from those terrible fates, as if somehow I passed through an ever-closing portal into my future, stumbled through it, like at the Polar Caves in New Hampshire, which my family visited one year, geological formations turned tacky tourist site, where we barely fit through the Lemon Squeeze, a narrow passage between rock walls, or the Orange Crush, a tunnel-like passage through granite, the opening a small bright hole to the other side.

When I hear about so many people addicted to opiates now, I wonder if that would have been me. I never heard of heroin or opiates in Walpole in the 1970s, but I took every drug that was available then — angel dust, acid, cocaine, crystal meth, speed, Quaaludes, pot, hash oil, amyl nitrate (sniffing from a small amber vial), nitrous oxide (whippets).

What pulled me off the path of self-destruction was a confluence of events, part fate, part choice. There were people who left me, people I left behind—Paula and Alison, my drug cohort. Weekly for a year my counselor, Jim, provided a guiding hand, a nudge toward healthier choices. The natural process of maturing helped, as I outgrew the neurobiology of the adolescent brain, less understood in the 1970s. Like 90 percent of errant youths, I "aged out" of delinquent behavior, which peaks at sixteen or seventeen, then declines.

The family support I had for the first twelve years of my childhood, before my parents split, before the system broke down, provided a foundation beneath me, even though I lost my footing for a while. It was like when I swam in that mile race, struggling to keep my head above water, swimming in a crooked line, but I kept going, and at the end of the race, at my weakest, it felt like solid ground rose up beneath me and I could stand. My family had been there for me even if I hadn't seen them. When I had no friends that summer, I hung around with Sally and her friends, or Joanne and hers. When I needed Sue after Nicky and I broke up, she drove four hours round trip. My father tried to help me; even if he didn't know what to say or do, even if he couldn't reach me, he never gave up. My mother showed me early on that help was available through counseling and arranged for me to see Jim when I asked.

Even Gerry Delaney, my misfit guide, helped by treating me as an equal, though I was four years younger, talking with me about books, stirring the embers of my intellect as I headed into my senior year. Gerry took me out of Walpole, showed me a different corner of the world.

All of these factors contributed, but one act was fundamental to resetting my course, like flipping a railroad switch to divert a train, something antithetical to my rebellious nature, that contradicted my mother's lessons about *fighting my own battles,* a step that can be most difficult for teenagers, who yearn for independence: I asked for help.

In 2016 I heard Boston's police commissioner, William Evans, on the radio talking about how juvenile arrests and crimes were down 21 percent that year, which he attributed to diverting kids from the criminal justice system by providing them with educational, vocational, and out-of-school opportunities. The program significantly reduced delinquent behavior, including a 59 percent reduction in carrying weapons, a 64 percent reduction in aggressive behavior, and a 71 percent reduction in victimization. Iceland, too, dramatically reduced teen delinquency by switching the rush kids got from drugs and crime to natural highs. Iceland's government set up after-school classes, sports programs, clubs for dance, arts, music, life-skills training — all free, with free transportation. Kids "do not need to use substances," said Inga Dóra Sigfúsdóttir of the University of Iceland, "because life is fun, and they have plenty to do."

These programs feel intuitively right to me. I was destined to rebel, to do or say the taboo thing; it's hardwired into my personality. In the fall of ninth grade, when my parents were still tracking my progress, my Spanish teacher complained about me on parent-teacher night: *Nov. 12, 1974: Miss Ripley told Dad I did anything I damn please in class.* Miss Ripley gave me extra credit for "outstanding achievement in Spanish" and allowed me to correct my classmates' quizzes, which I took literally, erasing wrong answers and writing in correct ones. At fourteen, *I did anything I damn please.*

If I had channeled my rebellious energy into activities I loved and that were meaningful to me — even if I didn't know how meaningful then, the seeds of who I'd later become, books and writing, swimming, nature and environmental conservation, political activism — I might have had a less destructive adolescence. I think of the hours that Paula and Alison and I dedicated to planning the B&E, the collaborative nature of our project, our perseverance, the determination to follow through, to take a risk — admirable qualities, misapplied.

I wonder if my teenage years might have been different if, instead of just driving by, I'd had the chance to go behind the walls of Walpole Prison, as some Walpole High School students did in 2016, to talk with inmates about their choices that led to jail time—the school using the prison as a behavior modification tool, as my mother had. If I'd met face-to-face with a young man trapped in that brutal place, if he'd said, "I made this choice to take drugs at a young age. I never thought I'd end up here," as he told those Walpole High School students—would I have listened?

## 9

## *Stop the Dust*

I NEVER MET ANYONE OUTSIDE WALPOLE WHO'D SMOKED ANGEL dust, even kids at UMass from tough towns like Revere and Dorchester. Because of that, I linked the drug problem in the town of Walpole to the prison. I thought the concentration of badness had leached into the community, like the time the cesspool in our backyard overflowed and our yard stank. I thought the prison had somehow attracted drugs to Walpole, even though I *knew* the dealers — Lighty and Wayne Kosinski, who'd been in my house, Paula's brother, Duane, who'd first turned me on to angel dust, and Paula, my friend. I thought the badness was contained behind the cement walls of the prison, but the badness was in us, too, the citizens of the town, its sons and daughters.

In 1978, a few months after we graduated from high school, Paula and Alison were arrested. Alison was charged with possession of a controlled substance with intent to distribute, and Paula was charged with possession of Class B, C, and D controlled substances and conspiracy to violate Massachusetts's controlled substances law. They were lucky they'd been caught a few months before President Carter signed the Psychotropic Substances Act that November, which upgraded angel dust from Class III to Class II (with cocaine and methamphetamine), and doubled jail time for dealing PCP from five years to ten. Maybe the charges were dropped, or maybe as first-time offenders Paula and Alison received suspended sentences, but they were both in college that fall, and I was glad for that outcome, glad they'd been given another chance.

I knew the angel dust dealers in Walpole, but I didn't know where they got their drugs until recently when I stumbled on the connection, the person who supplied Lighty and therefore the person who likely provided most of the angel dust I smoked. Jeff, Sue's high school boyfriend, who'd distributed dust for Lighty, told me not long ago that Lighty's source had been his "cousin in Rozzy" (Roslindale), a Boston neighborhood known for dust. Once when Nicky's sister, Andrea, and I couldn't find dust in Walpole, we took a bus to Rozzy and asked the first kid we saw for angel dust. He disappeared around a corner and returned with a couple of damp joints, which we smoked on the ride home.

In November of 1977 — two months after I'd quit dust for good — Paul Ragucci, a twenty-two-year-old man from West Roxbury, which bordered Rozzy, was arrested with twenty pounds of PCP, at that time the largest seizure of angel dust in Boston's history. Twenty pounds of PCP would have been 9,072 grams of angel dust, enough to supply nine dealers with over 1,000 grams each. The street value at $10 a gram was over $90,000, or about $356,000 today. Ragucci, it turns out, was Lighty's "cousin in Rozzy" — a one-man pipeline of PCP into Walpole.

Ragucci, a former high school all-scholastic athlete (he played fullback for Catholic Memorial High, an all-boys prep school), was charged with possession and intent to distribute, which carried a mandatory minimum sentence of five years in Walpole State Prison. But Ragucci's attorney exploited a Massachusetts statute that allowed people arrested for drugs to receive treatment instead of jail. If the arrestee failed to complete treatment, then the prison sentence would be invoked, like a layaway plan for punishment. Instead of going to jail, Ragucci went to rehab, even though the law specified it should apply only to drug charges "other than the sale or manufacture."

There's less forgiveness for PCP-related crimes when they're committed by minorities. About a decade after Ragucci's arrest, Garry

Jordan, a black man from Washington, D.C., also a former athlete (in 1978 he earned the highest scoring average for a freshman on Niagara College's basketball team), was sentenced to sixty years in prison for dealing PCP, an extraordinarily harsh sentence, especially since the amount he sold to an undercover cop, 500 grams, was twenty times *less* than the amount possessed by Paul Ragucci, who didn't spend a single day in prison.

That I was able to smoke angel dust for nearly two years without any-one confronting me — not my parents or teachers or guidance coun-selors, no cops, no adult saying, *Hey, that's a bad drug, don't take it* — might have been because of a lag time before police, drug agencies, and schools learned of PCP and understood the scale of its use. When the first cases of PCP overdoses appeared in emergency rooms in the mid-1970s, medical staff thought they were witnessing an epidemic of schizophrenia. A doctor in a D.C. hospital said that PCP had "no equal in its ability to produce psychoses nearly indistinguishable from schizophrenia."

The first article on angel dust in the *Boston Globe*, which my father read daily, was in the summer of 1977, when I was hanging around with Gerry Delaney, my dust use waning. "Hospital emergency rooms are ministering increasingly to users of an animal tranquilizer possessing a kick so bizarre it's been dubbed 'embalming fluid,'" the article said. In July 1977 the *Boston Globe* called PCP "a recently popularized drug" and reported a raid of "the first alleged PCP factory" in New England, in a Rhode Island town a half-hour from Walpole, where police seized $500,000 worth of PCP, worth about $2 million today.

That July a *Boston Globe* reporter immersed herself for a few days with teenagers in Revere, a rough town north of Boston that had a PCP problem, like Walpole. One Revere teenager said that angel dust had been around for a year or two and that "all of a sudden it came in and wiped out the whole town." Some of her friends, she said, "had no minds left." That was the name we called burnouts who smoked too much angel dust: no-minds.

Boston's Charlestown neighborhood launched a campaign called "Stop the Dust" and held a public meeting with Boston Bruins star Bobby Orr and former heavyweight boxer Tom McNeeley, the athletic director at Norfolk Prison. Charlestown's police sergeant told the packed audience, "Angel dust is not an upper, and it's not a downer. It's an inside-out drug."

In August 1977, *U.S. News & World Report* wrote that angel dust was reaching "epidemic proportions." Federal officials estimated that by late 1977 nearly seven million people, most of them aged twelve to twenty-five, had tried PCP. In 1978, a full year after I'd quit angel dust, the National Institute on Drug Abuse released a public service announcement about PCP, starring Robert Blake of *Baretta,* who warned, "Don't go near it. It's a rattlesnake and it will kill you." And Paul Newman and Joanne Woodward narrated a documentary called *Angel Death,* which showed a monkey, a jaguar, and a fifteen-year-old boy high on PCP. The boy, George, snuck away to get dusted, and the cinematographer caught him on film, glassy-eyed, slack-jawed, unable to answer simple questions.

By late 1977, Walpole police admitted that PCP was "one of the department's largest enforcement problems." Shortly after, though, another article reported that the cops had "broken the back of the angel dust nest in town" and had arrested every major dealer. The claim was optimistic. In 1980, Walpole police, state police, and the feds busted a "major drug factory" in Sharon, a town adjacent to Walpole, involving a Walpole man, Paul Papasodero. Cops found enough gallons of liquid PCP to make $3 to $5 million worth of angel dust ($9 to $14 million today) and a pound of finished angel dust with a street value of $350,000 then.

For two years a PCP ring operated out of Medfield State Hospital, orchestrated by Walpole prison inmates working there as trusties. The inmates forged state purchase orders and procured hundreds of thousands of dollars' worth of PCP chemicals, at state expense, which they had drop-shipped to five angel dust labs in Massachusetts. In 1981 twenty-one men were charged, including the trusties working at Medfield, a few men on parole from Walpole Prison, and two men still inside Walpole Prison.

Angel dust lingered in Walpole for more than a decade after I quit, into the early 1990s. I last saw Ian O'Shaughnessy in the fall of 1978, after we'd graduated from high school—this boy who, at twelve, had given me my first kiss. He was probably nineteen when I saw him outside the A&P carrying a bag of diapers for the baby he and his girlfriend had just had, proud to be a father. We talked briefly, his voice still raspy as I remembered it from sixth grade, and those big blue bedroom eyes. I remember that summer day we kissed in the parking lot of Fisher Elementary, how he pressed his warm, dry, chapped lips against mine—I can almost feel the sandpaper texture—and held them there unmoving for a few seconds. I remember his scent, like dried sweat, unwashed hair. A few months after I ran into him in 1978, he was arrested for possession of angel dust with intent to distribute, and nine years later, in 1987, he was arrested again for possession of PCP.

And Ken Coffey, my friend in woodworking my senior year of high school, whom I'd ask for help since I was embarrassed to ask the teacher. Ken had a pumpkin face and a big yucking laugh, and we laughed a lot in that class, especially when I pressed too hard on my warped checkerboard drying in vise grips and the tension made the glued-up board explode into forty-eight individual blocks that sprang all over the room, and we doubled over laughing because Ken had made the terrible suggestion that I push down on the warp. Five years out of high school Ken was caught with angel dust, and then again two years later, in 1985, that time winding up in the Dedham House of Correction, the county jail, for a year.

Angel dust is back. Or, more accurately, it never left. The drug moved into predominantly urban areas with mostly minority populations and now gets little media attention, not the way it did in the 1970s and early

1980s when it invaded white suburbs. PCP garners only local headlines, which are frequent, though they represent only arrests associated with the drug, not its pervasive use. In the hundreds of articles I reviewed over three years, about 90 percent showed mug shots of African Americans; the remainder were Hispanics or an occasional white person.

PCP is now called "wet" and is sold as a tincture, a more potent form of the drug, into which people dip cigarettes or pot. The Drug Abuse Warning Network reported a "sharp spike" in PCP use, with PCP-related emergency room visits rising 400 percent between 2005 and 2011. In 2012, Los Angeles police made the largest bust of PCP in history, seizing $100 million worth. They confiscated 130 gallons of finished PCP and another 1,000 gallons of precursor chemicals to manufacture the drug.

Even when PCP ruins the lives of famous people, its use barely registers against the larger tragedy of the opioid crisis. In August 2016, E'Dena Hines, the thirty-three-year-old granddaughter of actor Morgan Freeman, was stabbed to death by her boyfriend, Lamar Davenport, while he was high on a "bad batch" of PCP. Three months later, in November 2016, Christopher Barry, the thirty-four-year-old son of former Washington, D.C., mayor Marion Barry, died of "acute phencyclidine toxicity," after struggling for years with PCP addiction.

Perhaps the best-known case was Aaron Hernandez, former tight end for the New England Patriots. In 2013, not long after Hernandez signed a $40 million contract with the team, he shot to death his friend Odin Lloyd. A family friend told *Rolling Stone* magazine that Hernandez had been "twisted on dust now for more than a year." Another acquaintance said Hernandez "was regularly high or out of his mind on angel dust." He called Hernandez a "d-head."

Hernandez played football at the Patriots' home field, Gillette Stadium, where in ninth grade, when it was called Schaefer Stadium, I sold hot dogs in the concession stands, where one Sunday I lingered after a game ended to meet players as they strolled out of the locker room, excitedly collecting autographs from the young quarterback, Jim Plunkett, and the wide receiver, Randy Vataha. From his glory on the grid, like in a Greek tragedy, Hernandez landed in Walpole Prison, two miles from the stadium, where he served the first few months of a life sentence before he was transferred to Souza-Baranowski, a newer maximum-security prison, in Shirley, Massachusetts. In 2017, Hernandez hanged himself in his prison cell.

10

*Body Leaping Backward*

TEN YEARS AFTER I GRADUATED FROM HIGH SCHOOL, I TRAVELED home to Walpole from Michigan, where I'd moved after college, to visit my family for the holidays. That Christmas we were all grown up. Sue was twenty-nine, Sally twenty-eight, I was twenty-seven, Joanne twenty-six, Patrick twenty-four, Barbie twenty-three, and Mikey seventeen. My mother was still with Ed, my father in a serious relationship, his first since he'd separated from my mother fifteen years earlier.

Like me, Sue and Joanne had graduated from UMass — the three of us were there together one year, meeting at Sue's apartment for Sunday dinners. Sue was married and had given birth to her first child a few days after finishing her MBA. Joanne was engaged, working in human resources, and I worked for an environmental nonprofit, a job I loved. After living in a tent on Cape Cod and cooking in a four-star restaurant, Chillingsworth, Sally moved to Michigan near me, finished her degree in art, met her future husband, Terry, then moved back East. Patrick went west, pushing farther each time, to Nevada, Oregon, California, finally Hawaii, where he earned his degree in criminal justice and started a Roller blading league for four hundred kids in Honolulu. Barbie had just graduated with honors from Boston College, and Mike was in his last year of high school. He worked part-time at a gas station on Route 1A, fielding questions on weekends as I had a decade earlier, drivers asking, "How do you get to Walpole Prison?" even though it wasn't called Walpole Prison anymore.

After the tumultuous 1970s, when the prison made the daily news, in the early 1980s a group of Walpole residents campaigned to separate the town from the prison. They were tired of the town being "tainted" by "rapists, murderers, and hatchet killers" sent to "Walpole," annoyed by newspaper headlines like "Child Molester Sent to Walpole." One idea was to print bumper stickers: *Yes. You Can Leave Walpole.* In the end they decided to rename the prison. The town's representative got a bill passed in the legislature, and the group held a "Name the Prison" contest in town, which drew some six hundred entries, including Massachusetts School for the Misguided, Inmate Inn, House of Hope, the Zoo. The winner was Cedar Junction, after a defunct railroad station near the prison. Cedar Junction, as if it were one of the condo complexes springing up in Walpole during the 1980s development boom.

The prison name was changed, a supermax unit was added in the 1990s, and then the prison itself was slowly hidden from view. A new, longer, winding driveway was cut through the woods, and trees and vegetation grew to obscure the old entrance, where we used to park for the Hobby Shop. Driving along Route 1A, you can no longer see the imposing white walls, as if by hiding the prison you hide the problem of crime and incarceration. Across from the prison, surrounded by weeds, is the small brick shelter where visitors or newly released ex-cons waited for the Greyhound to take them out of Walpole.

On Christmas Eve that year, we gathered in the living room, my father, too, and told stories of embarrassing moments, family lore — how Joanne nearly crashed into our house when she was learning to drive; the time Patrick threw a dart at me in the Orange Room and it stuck in my thigh and we doubled over laughing; the time when Sue was drunk and put Cheez Doodles in her ears; the time my mother charged into the Orange Room and yanked the record off the stereo because she thought Joanne and Barbie were playing Charles Manson music, "Helter Skelter," when it was Neil Young singing, *Helpless helpless heelpless,* my mother and father so helpless to help us back then.

We laughed at these hilarious moments, the wild years. We thought the pain of our past was behind us, that we'd outgrown our youthful troubles. Then Barbie shared a story about me, one I hadn't remembered until she jarred it from the dusty archives of my mind. My friend Alison and I were sitting on our screened-in porch rocking back and forth, rocking and rocking. We'd just smoked angel dust and we couldn't stop rocking, but we weren't sitting in rocking chairs. Alison and I sat cross-legged on the wooden floor, rhythmically rocking. Barbie, who was eleven, was unnerved by our repetitive, nonsensical, metronomic, idiotic motion. She told us to stop, but we couldn't hear her, so we rocked and rocked, and Barbie became so disturbed that she left the room.

After Barbie told this story, nobody laughed. My father said, "Jesus," then there was an uncomfortable silence. This was not a funny sowing-wild-oats story. This was a story of two fifteen-year-old girls incapacitated by a drug that rendered them unable to think or function. After that the party broke up and we all went to bed. I felt sick with embarrassment and shame that I'd been rocking like the kids I used to see when I swam at the Wrentham State School — the girl who'd bragged about her straight A report card, who'd scored in the high-90s percentiles in the Iowa Tests, that promising girl with the smart mouth.

Upstairs in my old childhood bedroom, lying on that single bed that used to fly out the window when I came home dusted, I sobbed, a pillow stuffed in my mouth so no one could hear me, the walls thin in that small house.

The following year my parents listed our house in Walpole for sale. Mikey had turned eighteen, and so, according to my parents' divorce decree, the house could be sold. My father wanted his half of the money for a new house with his soon-to-be new wife. Years earlier the DEAD END sign at the top of our street had been changed to NOT A THROUGH WAY, which someone graffitied to read NOT A T OUGH WAY, and now it was changed again to CUL-DE-SAC, a fancier term reflecting the town's rising real estate values. I saw the FOR SALE sign staked in our front yard, but I didn't believe the house I grew up in would be sold. I thought one of my siblings would buy it, or that we'd band together to save the house.

While I was a thousand miles away in Michigan, my mother began to deaccession the museum of my childhood. She donated a dozen boxes of books to the Walpole library, furniture to charity, twenty bags of clothes to Wrentham State School. She held a yard sale but neglected to tell my siblings in Massachusetts. Sue, who lived an hour away, learned of the sale at the last minute. "By the time I got there almost everything was gone, so I quickly grabbed the fondue pot," she said. We laughed. She never uses the pot, but she keeps it all the same.

That fondue pot, like a magic genie lamp, summons a vision: our dining room table laden with dips and sauces and cubes of meat we skewered and plunged into boiling oil. When I was a girl, a fondue for dinner was adventure enough to make the whole day juicy with anticipation, was all it took for happiness. When my mother went back to work, there wasn't time for such frivolity. By the time Mikey was eight, often alone after school, he'd cook himself cans of Campbell's soup for dinner, thick and pasty. Nobody told him he was supposed to add water.

After most of my siblings moved away, the pool was neglected. The water turned green, then brown, the lining sagged and slipped bottomward. After the yard sale, my mother and Mikey rented a Bobcat front-end loader and plowed everything that remained into the hopper of the pool, including the pool itself—the wobbly aluminum walls, the filter, the liner, the rotting deck—my mother undoing in a single day what we'd built that long, hot, chain-gang summer. They filled that huge grave with loam, smoothed topsoil over it, and sprinkled grass seed. It seemed fitting that our swimming pool, constructed against all good professional advice, paid for with "a wing and a prayer," as my mother said, built of our sweat and muscle and desire, and the temporal marker of the departure of my father from my daily life, was collapsed.

In August my mother and father met at the house for the closing. I imagine my mother that day looking out the window at the Gibsons' lawn, which was still patchy and bald. Every Easter and Thanksgiving as we sat around the dining room table, my mother, with a view out the window, would say, "There's Arthur Gibson mowing his lawn." She'd shake her head, because who mows his lawn on a holiday? "Something is not right in that house," she'd say, but all along something had been not right in our house. Standing in the driveway after the closing, my mother and father hugged each other and cried.

As children, we want to believe that our parents created us out of love, that love existed, at least for a while. One night nearly twenty years after my parents split, we watched family movies in my father's new house, the super-8 films he transferred to video. As I watched them for the first time as an adult, I saw something that surprised me — my mother and father in their early years, fond and affectionate. In one shaky silent clip my mother stands on a ladder to paint the trim on our first house, in East Walpole. She's wearing Bermuda shorts, one knee jauntily lifted as she reaches with her brush, my father slowly panning, caressing my mother with the camera until she turns and smiles, catches him in the act, which we all see now, his appreciating eye surveying her form, her beauty. She dabs the paintbrush in the air toward him, winks.

Later in the clip, my father is on the ladder as my mother films. He turns to smile at her. I can see what was characteristic of him, what I came to know — that he was ineffectual at any type of home repair, ill-suited for ladder work, out of his element, in spite of his athleticism (a college football injury kept him out of the Korean War). But he is smiling and young and handsome and happy.

In another snippet that lasts no longer than fifteen seconds there's a crowd at an ice rink, the camera jerky and rapid and then a brief clear focus as skaters glide toward and past the camera in carousel fashion. Soon from the crowd I pick out two slender, graceful skaters, a dark-haired, dark-eyed beauty in stirrup pants tucked into white figure skates, the white short-waist leather jacket that hung in our closet when I was a child but that I never saw my mother wear; her tall, handsome, black-haired partner holding her hand as they effortlessly glide toward us, my mother and father now recognizable but also not recognizable as the parents I knew, holding hands, smiling, in sync with each other, here in front of me, now rounding the corner, gone.


After the meeting when my parents told us that they were separating, nobody in my family mentioned it again. "The kids will bounce back," my parents' marriage counselor assured them, the conventional wisdom of the day, but to bounce back, you must hit bottom first. I never knew the point at which my parents' separation became final. They never announced that the "trial" had failed and they were divorcing, not the way they'd sat us down to announce the separation. I never heard about my parents going to court, signing papers. It wasn't until I was in my forties that I learned that my parents had tried to reunite a third time (they'd tried right after the separation), a decade later, when I was in college. They must have realized they'd lost something special. But they couldn't go back; you can't go back and fix things, only go back to understand them.

I didn't know for years that my father was offered a huge promotion to head the London office of his company, but he turned it down because he didn't want to be far away from his kids. I didn't know that the bout of drinking and womanizing after the divorce was his way of escaping loneliness and sorrow; to me it looked like he was just having a good time. Years and years after my father left, Sue told me that he'd been suicidal, that one day he found himself driving faster and faster, wishing, willing, to end his life in a car accident. It took decades — when my father was in his sixties and I was in my forties — for him to tell me how he struggled back then, how one afternoon when he tried to run a meeting, his boss recognized that my father was a wreck, was coming undone. *Go home,* his boss kindly said, *take the day off.*

Only then did I begin to comprehend my father's loss, imagine the silence that greeted him when he came home from work, no gaggle of children racing up to him and clinging to his legs, no piano to play, no scent of roasting chicken, no noisy kids dancing around the living room or playing games in the yard. Just the bare wood floors of the Savin Hill flat, the dingy secondhand day bed, the spare cupboards, stray condiments in the refrigerator, the empty hours each night after work, the paralyzing quiet.

After my father moved out, Joanne recorded a greeting for him on his answering machine, which he kept until he remarried, at fifty-six. I asked him once why he didn't update his answering machine greeting. He said he loved to hear Joanne's voice when it played, couldn't bear to erase it. When he was eighty and in the hospital dying of cancer, dehydrated, as I brought my father a glass of water I reminded him of our chants for *cold, cold, cold water* each night when he tucked us into bed. "That was the joy of my life," he said. I felt then, as I never fully had, how painful it must have been for him to lose those bonds with his children — that he enjoyed being a father and was a good father until our family fell apart and he didn't know anymore how to be a father.


For years after my mother buried everything in the pool, I had dreams of digging in the side yard as if it were an archeological site. In the dreams it was always nighttime, as if I were a grave robber, the dream always interrupted by a light that flashes on, or by someone calling my name, as if I were trespassing on someone else's property. For years I wondered what my mother had plowed into the hole. My love letter from Ian O'Shaughnessy in sixth grade that began, "I'm sorry I threw the football at you"? My black patent-leather Mary Janes from fourth grade, with the square toe and oversized buckle that made me feel like a pilgrim? I loved those shoes so much that I cried when I left one at my grandmother's in New York. "Mom, *please*. Make Nanny send it." A year later my grandmother mailed the shoe. When I opened the box, I couldn't imagine why I'd been so attached to those stupid loafers, but I kept it. I wanted that shoe to teach me something, to remind me how to be excited about a thing as simple and inconsequential as a shoe — how to be happy, really.

What else went into that hole, that hopper that was filled with clear, cold water that I smashed into, water that broke the fall of my body leaping backward? Gifts we made for Mother's or Father's Day: rock paperweights, ceramic pencil holders? An entire set of Funk and Wag-nalls encyclopedias — Aardvark to Zululand? Sally's Visible Woman set? My ESP cards with the shapes and lines you were supposed to guess through concentrating, proving you could read minds? Patrick's Batman lamp from the prison Hobby Shop — is that in the hole? What about stuff we had when we were teenagers: my brown-and-orange cheerleading skirt from West Junior High, clogs, the psychedelic go-go girl lamp from the Orange Room? My single swimming trophy? That must be in the hole; otherwise, where is it?

One day years after the house was sold, Joanne and Barbie ran into each other in front of it. Neither lived in Walpole anymore; both had detoured from wherever they were going at the same moment to drive by the house. I didn't visit the house for nearly thirty years. I didn't want to see the weeping willow, no taller than my mother when she planted it, now grown above the roofline, or the side yard where no space-picket fence guarded no wobbly aluminum pool. I didn't want to see the clapboards painted pumpkin or sage by the new owners, rooms added on like bastards.

In college, when my conscience returned from its period of absentia in high school, I dreamed of paying back Mr. Barnes the few hundred dollars Paula and I had stolen. But I never did, perhaps for the same reason I rarely went back to Walpole once I'd moved away at twenty: there was too much to forget and avoid. Making one small reparation meant thinking about or acting on so many others that I could not correct; I could not possibly repair all that I'd stolen, damaged, hurt.

Not long ago I asked my mother why she'd stolen. Her eyes were downcast, and she was uncharacteristically quiet. "I had a lot of children and no money," she said. She was in her late seventies, just retired from forty years of working as a nurse. "That's sad," she said. "I was so honest." My mother told me she regretted using our childhood savings for the pool. "When I saw you all had no money for college, I felt bad." I told her we loved that pool. That pool was more than a respite on hot summer days. That pool demonstrated to me what a woman not even five feet tall in the world, a single woman with a high school education and seven kids, could accomplish.

After my parents sold the house in Walpole my mother bought a condominium, which was all she could afford. All the units were exactly alike, except the interior color schemes were either beige or blue. Mikey called the place Beige Number Nine. The first time I visited Beige Number Nine, my mother handed me a manila envelope stuffed with photos she'd divided among all of us. She'd cut up group shots, a weird reenactment of the splintering of my family. My envelope contained mostly pictures of myself, but in each of our packets she included one or two photos of all of us. She must have run out of family shots, so on some photos she taped in whoever was missing, most often my father. In one photo she gave me, the misaligned curtains are a dead giveaway — that, and Patrick's enormously long arm around the shoulders of Sally and my patched-in father.

Along with the photos, my mother gave me a single cardboard box, the remains of my childhood, which contained my photo albums, college texts, my high school yearbook, and the walking stick I'd carried home from California, my handwritten note still taped to it a decade later: "Do not ever throw this stick away." I must have known to guard against loss. The box also contained my diaries, which I'd forgotten about, a yellow five-year diary and the diary from 1975, when I first smoked angel dust. The gilt-edged pages were too small to contain all that happened on some days, so I continued the entry on an earlier page I'd left blank, the diary moving forward in time and then backward. Reading those entries, I had a sense of time unwinding, the future written in the past.

Of the few things my mother saved from burial in the pool, the diaries are most precious to me, for they returned years of my life I'd mostly forgotten. When I first read the pages from fall 1975 when I began to smoke angel dust, I wept for the loss of myself, pitied myself, was once again self-centered. I thought those events were so far in the past that I would not be affected. But I *felt* the memories — not body or intellect, but a soul memory, a heart memory. I felt unhinged and weird, like a tormented teenager, flooded with a sense of alienation that left me off-kilter for days, unable to be around people. Reading the diary triggered remnant feelings, flipped on that switch, the current weaker and sputtering but felt.

In revisiting my adolescence, at first I felt enraged with that fifteen-year-old girl. I wanted to shake her roughly, slap her face, grab her by the collar like my mother did that night when I didn't come home. I wanted to wake her up, knock some sense into that girl, penetrate that faux-tough veneer. But I recognized that she didn't need tough love, just love, for someone to speak kindly to her, to help her find her way, to tell her she was smart and strong, remind her she was a good person, or could be, would be, to give her permission to speak up, speak out, speak her mind.

Recently when I was talking to Patrick about our childhood, he said, "You always stuck up for me." And in my forties, when I was helping my mother fight for her rights in a lawsuit, when I challenged and wrote letters and spoke on her behalf, she said one day, "All those qualities we punished you for? I'm glad you have them." *Loudmouth. Smart aleck. Back-talker. Fresh.*

Instead of anger, I try to see that girl in the diary with compassion, that self-destructive, lost, misguided girl. I try to understand her, and girls like her, to forgive her.

## Author's Note

This story depicts particular events at a particular moment; it is not meant to be representative of the town or people of Walpole. I have used real names for my family, public figures, people I interviewed who gave me permission, and the deceased. To protect the privacy of others, I have changed names, other identifying details, specific locations, and in a few instances I have conflated characters to assure privacy.

A short section of this memoir draws on material from an essay, "Body Leaping Backward," published in *Fugue Literary Journal*.

# Acknowledgments

I'm enormously grateful to the people who believed in this book, who midwifed it into the world: Rayhané Sanders at Massie & McQuilkin, so sharp and so mighty; Deanne Urmy, the smartest, kindest editor a writer can hope for; and the excellent team at Houghton Mifflin Harcourt: Jenny Xu, Megan Wilson, Liz Anderson, and Lisa Glover.

I'm grateful to the University of Massachusetts Lowell for time off to write, to my colleagues in the English Department for support and encouragement, and to my fellow creative writers for inspiring me: Maggie Dietz, Andre Dubus III, and Sandra Lim. Thanks to Andre in particular for benediction as I sent this story out into the world.

George Hart, the director of libraries at UMass Lowell, went above and beyond to provide me with access to research databases. I'm grateful to the Walpole Public Library for its excellent digitized archives; I'm happy to see that my old hometown has a great library.

Ms. Dorothy Gill generously shared with me her experiences as a social worker in Walpole Prison. I'm ever grateful to Jeff Day for his willingness to share his memories of Walpole in the 1970s, and for being brave, honest, and tenderhearted; rest in peace, my friend.

Thanks to these dear friends, upon whom I've prevailed again and again to read pages, to shore me up, calm me down, egg me on, who listen to my complaints, laugh at my jokes, and tell me the truth. I am a better writer, a better person, for these generous souls: Jason C.

Anthony, Jennifer Cognard-Black, Heather Hardy, E. J. Levy, Sandra Miller, and Nancy Sferra.

My sisters and brothers, to whom I am forever grateful — what can I say. You let me — helped me — write this book, and I can see no other reason why but love. How lucky I am: thank you. Thanks most of all to my mother.

# Sources

This memoir draws on the following sources (selected):

*Boston Globe* archives 1950–1995
*Walpole Times* archives 1950–1995

Agar, Michael H., George M. Beschner, and Harvey W. Feldman. *Angel Dust: An Ethnographic Study of PCP Users.* Lexington, MA: Lexington Books, 1979.

Bailey, Beth, and David Farber, eds. *America in the Seventies.* Lawrence: University Press of Kansas, 2004.

Baum, Dan. *Smoke and Mirrors: The War on Drugs and The Politics of Failure.* Boston: Little, Brown, 1996.

Bernstein, Nell. *Burning Down the House: The End of Juvenile Prison.* New York: New Press, 2014.

Bissonette, Jamie. *When the Prisoners Ran Walpole: A True Story in the Movement for Prison Abolition.* Cambridge, MA: South End Press, 2008.

Bondi, Vincent. *American Decades: 1970–1979.* Farmington Hills, MI: Gale Research, 1995.

Borges, Ron, and Paul Solotaroff. "Aaron Hernandez: Inside the Dark, Tragic Life of a Former Patriot Star." *Rolling Stone,* August 28, 2013, www.rollingstone.com/feature/the-gangster-in-the-huddle.

Boston Public Radio. Margery Eagan and Jim Braude, hosts, interview with William Evans, Boston police commissioner, November 27, 2016.

Burrough, Bryan. *Days of Rage: America's Radical Underground, the FBI, and the Forgotten Age of Revolutionary Violence.* New York: Penguin, 2016.

Carroll, Peter N. *It Seemed Like Nothing Happened: America in the 1970s.* Newark, NJ: Rutgers University Press, 1982.

Chesney-Linda, Meda, and Randall G. Sheldon. *Girls, Delinquency, and Juvenile Justice.* Boston: Wiley Blackwell, 2014.

Clouet, Doris H. *Phencyclidine: An Update.* NIDA [National Institute of Drug Abuse/Department of Health and Human Services] Monograph 64, 1986.

Cohen, Bertram D., Gerald Rosenbaum, Elliot Luby, and Jacques Gottlieb. "Comparison of Phencyclidine Hydrochloride (Sernyl) with Other Drugs." *Archives of General Psychiatry* 6 (1962): 395–401.

Cosgrove, Judith, and Terry G. Newell. "Recovery of Neuropsychological Functions During Reduction in Use of Phencyclidine." *Journal of Clinical Psychology* 47, no. 1 (1991): 159–68.

Cowie, Jefferson. *Stayin' Alive: The 1970s and the Last Days of the Working Class.* New York: New Press, 2010.

Dellelo, Robert, and Christopher Lordan. *The Factory: A Journey Through the Prison Industrial Complex.* CreateSpace Independent Publishing Platform, 2016.

Dynda, Russel S., Warren Jamison, and Michael McLaughlin. *Screw: The Truth About Walpole State Prison by the Guard Who Lived It.* Far Hills, NJ: New Horizon, 1989.

Edelstein, Andrew J., and Kevin McDonough. *The Seventies: From Hot Pants to Hot Tubs.* New York: Dutton, 1990.

Farber, M. A. "Veterans Still Fight Vietnam Drug Habits." *New York Times,* June 2, 1974, https://nyti.ms/1ku85LS.

Finn, Jessica. "Harrowing Final Hours of Morgan Freeman's Granddaughter's Life." *Daily Mail* [UK], April 12, 2018, www.dailymail.co.uk/news/article-5605623/Morgan-Freemans-daughter-spent-day-granddaughter-fatal-stabbing-boyfriend.html.

Frank, Gerold. *The Boston Strangler.* New York: Signet, 1966.

Frum, David. *How We Got Here: The 70s, the Decade That Brought You Modern Life — For Better or Worse.* New York: Basic Books, 2000.

Hamm, Duane C. *Manumission: The Liberated Consciousness of a Prison(er) Abolitionist.* XLibris, 2012.

Jaffe, Harry. "The Tumultuous Life and Lonely Death of Marion Barry's Only Son." *The Washingtonian,* January 8, 2017, www.washingtonian.com/2017/01/08/the-tumultuous-life-and-lonely-death-of-marion-barrys-only-son-christopher.

Janos, Adam. "G.I.s' Drug Use in Vietnam Soared — with Their Commanders' Help." History Channel, "History Stories," April 18, 2018, www.history.com/news/drug-use-in-vietnam.

Jenkins, Philip. *Decade of Nightmares: The End of the Sixties and the Making of Eighties America.* New York: Oxford University Press, 2006.

Jenson, Frances E., and Amy Ellis Nutt. *The Teenage Brain: A Neuroscientist's Survival Guide to Raising Adolescents and Young Adults.* New York: Harper, 2015.

Kamienski, Lukasz. "The Drugs That Built a Super Soldier." *The Atlantic,* April 8,

2016, www.theatlantic.com/health/archive/2016/04/the-drugs-that-built-a-su
per-soldier/477183/.

Kauffman, Kelsey. *Prison Officers and Their World*. Cambridge, MA: Harvard University Press, 1988.

Kelly, Susan. *The Boston Stranglers*. New York: Pinnacle, 1995.

Levine, Elana. *Wallowing in Sex: The New Sexual Culture of 1970s American Television*. Durham, NC: Duke University Press, 2007.

Linder, Ronald L., Steve E. Lerner, and R. Stanley Burns. *PCP: The Devil's Dust*. Belmont, CA: Wadsworth, 1981.

Malcolm X and Alex Haley. *The Autobiography of Malcolm X*. New York: Ballantine, 1964.

Manning, Tom. "Tom Manning, A Short Biography." Internet Archive Wayback Machine, www.geocities.com/CapitolHill/Parliament/3400/tom-bio.htm.

Miller, Jerome. *Last One Over the Wall*. Columbus: Ohio State University Press, 1998.

Rae, George William. *Confessions of the Boston Strangler*. New York: Pyramid, 1967.

Remick, Peter. *In Constant Fear: The Brutal, True Story of Life Within the Walls of the Notorious Walpole State Prison*. New York: Reader's Digest, 1975.

Schulman, Bruce J. *The Seventies: The Great Shift in American Culture, Society, and Politics*. New York: Da Capo, 2001.

Sexton, Anne. *Anne Sexton: A Self-Portrait in Letters*. Boston: Mariner, 2004.

Shteir, Rachel. *The Steal: A Cultural History of Shoplifting*. New York: Penguin, 2011.

Smith, R. Jeffrey. "Congress Considers Bill to Control Angel Dust." *Science* 200 (June 30, 1978): 1463–66.

Strass, Todd. *Angel Dust Blues*. New York: Coward, McCann & Geoghegan, 1979.

Taylor, Jacob. "PCP in the American Media: The Social Response to a Forgotten Drug." Master's thesis, Department of History, Concordia University, Montreal, January 2011.

Torgoff, Martin. *Can't Find My Way Home: America in the Great Stoned Age, 1945–2000*. New York: Simon & Schuster, 2004.

*United States of America v. Garry Jordan*. U.S. Court of Appeals for the District of Columbia Circuit 810 F.2d 262, January 1987.

U.S. House of Representatives, Select Committee on Narcotics Abuse and Control. *Abuse of Dangerous Licit and Illicit Drugs — Psychotropics, Phencyclidine (PCP), and Talwin*: Hearings. 95th Congress, 2nd sess., August 8 and 10; September 19; and October 6, 1978. Washington, D.C.: U.S. Government Printing Office, 1979.

Walsh, Ryan H. *Astral Weeks: A Secret History of 1968*. New York: Penguin, 2018.

Wride, Nancy. "Return to Dust: Bane of the '70s, PCP Now a Supporting Player in the Saga of Aaron Hernandez." *Elements Behavioral Health*, December 7,

2013, www.elementsbehavioralhealth.com/addiction/aaron-hernandez-pcp
-making-comeback/.

Young, Emma. "How Iceland Got Teens to Say No to Drugs." *The Atlantic*, January 19, 2017, www.theatlantic.com/health/archive/2017/01/teens-drugs-ice
land/513668/.

YouthConnect Program, Boys & Girls Clubs of Boston. "Proven Results," www
.bgcb.org/what-we-do/youthconnect/, 3/15/2017.

# *Appendix*

Possible charges by legal definition, and penalties, for crimes committed (or accessory to) by me, my family, my friends* (selected)

## Massachusetts Law, Chapter 90

Section 10 — Operate motor vehicle without license [misdemeanor, 1 year]

Section 24(1)(a)(1) — Operate motor vehicle under influence of liquor/drugs [misdemeanor, max. 2.5 years]

Section 24(2)(a) — Unauthorized use of motor vehicle [misdemeanor, max. 2 years]

Section 24(2)(a) — Reckless operation of motor vehicle [misdemeanor, 2 weeks to 2 years]

Section 24(2)(a) — Leave scene of property damage [misdemeanor, 2 weeks to 2 years]

Section 24(2)(a) — False statement in application for registration [misdemeanor, 2 weeks to max. 2 years]

Section 24B — Possess false or stolen, misuse or forge registered motor vehicle document [felony, 2–5 years]

Section 32A — Falsify title certificate motor vehicle / unlawful possession altered title / false statement in application for title [felony, 2 to 5 years]

---

* Laws and penalties may have been different in the 1970s; these are current. *Sources:* Massachusetts Sentencing Commission Felony and Misdemeanor Master Crime List 2015, www.mass.gov/courts/docs/admin/sentcomm/mastercrimelist.pdf; also www.mass.gov/courts/selfhelp/criminal-law/misdemeanors-felonies. Massachusetts General Laws, Part IV, Crimes, Punishments and Proceedings in Criminal Cases, https://malegislature.gov/Laws/GeneralLaws/PartIV/TitleI.

## Massachusetts Law, Chapter 94C

Section 32(a) — Distribute or possess with intent, Class B [felony, 2.5 to 10 years]

Section 32B(a) — Distribute or possess with intent, Class C [felony, 2.5 to 5 years]

Section 32C(a) — Distribute or possess with intent, Class D [misdemeanor, 2 years]

Section 32A(a) — Phencyclidine, distribute or possess with intent [felony, 2.5–10 years, mandatory 1 year]

Section 32I — Drug paraphernalia, distribute, possess with intent, or possess [misdemeanor, 1–2 years]

Section 32J — Mfg./Dist./Dispense Class B or C substance with intent to distribute within 1000 ft. of school [felony, 2.5–15 years]

Section 32K — Drug, induce minor to possess [felony, 5 to 15 years]

Section 32K — Drug, induce minor to distribute [felony, max. 15 years]

Section 34 — Illegal possession Class B substance [misdemeanor, max. 1 year]

Section 34 — Illegal possession Class C substance [misdemeanor, max. 1 year]

Section 34 — Illegal possession Class D substance [misdemeanor, max. 1 year]

Section 34 — Illegal possession Class E substance [misdemeanor, max. 6 months]

## Massachusetts Law, Chapter 159

Section 103 — Damage railroad car [misdemeanor, max. 2 years]

## Massachusetts Law, Chapter 266

Section 5A — Attempt to burn motor vehicle [felony, 2.5–10 years]

Section 7 — Woods; wanton or reckless injury or destruction by fire [misdemeanor, 2 years]

Section 8 — Set fire on land [misdemeanor, 2 years]

Section 9 — Failure to extinguish fire on land [misdemeanor, 1 month]

Section 10 — Burning motor vehicle to defraud insurer, or attempt [felony, 2.5–5 years]

Section 16 — Breaking and entering at night [felony, max. 2.5 years]

Section 16A — B&E for misdemeanor [misdemeanor, 6 months]

Section 19 — Railroad car; breaking and entering [felony, max. 10 years]

Section 20 — Railroad car; larceny from, under $10,000* [felony, 2–5 years]

Section 27A — Motor vehicle or trailer; removal or concealment to defraud insurer [felony, 2.5 years]

Section 28 — Larceny, motor vehicle or trailer / receive/buy stolen motor vehicle [felony, 2.5 years]

Section 28(a) — Motor vehicle, malicious damage [felony, 2.5–15 years]

---

* 1975 values adjusted to current dollars

Section 28(b) — Conceal theft of motor vehicle [felony, 2.5–10 years]

Section 30A — Shoplifting by concealing merchandise, over $100 [misdemeanor, max. 2.5 years]

Section 30(1) — Larceny under $250 [misdemeanor, 1 year]

Section 30(1) — Larceny over $250 (under $10,000) [felony, 2–5 years]

Section 30(5) — Larceny from elder, under $250 [misdemeanor 2.5 years]

Section 60 — Receive/aid in concealment of stolen goods under $250 [misdemeanor, 2.5 years]

Section 60 — Receive stolen or false-traded property over $250 (under $10,000) [felony, 2.5–5 years]

Section 98 — Malicious damage, school property / malicious injury school building / vandalize school or church [misdemeanor, 2 years]

Section 102(c) — Explosives, possess [felony, 10–20 years]

Section 104 — Destruction of property > $250 [misdemeanor, 2.5 months]

Section 111A — Prepare or present false insurance claim [felony, 2.5–5 years]

Section 111B — False insurance claim, motor vehicle [felony, 2.5–5 years]

Section 127 — Destruction of property under $250 [misdemeanor, 2.5 months]

Section 127 — Wanton destruction of personal or real property over $250 [misdemeanor, 2.5 years]

Section 127 — Malicious destruction of personal or real property over $250 [felony, 2.5–10 years]

## Massachusetts Law, Chapter 268

Section 28 — Deliver drugs to prisoner [felony, max. 5 years]

Section 39 — Motor vehicle theft, false report of [misdemeanor, max. 2 years]